NOBEL

Rediscovering Nobility in the Age of Artificial Intelligence

Pershia Dashti

Dedication

To those who choose meaning over metrics, conscience over convenience, and truth over trends—

This is for the quiet nobility in every act of intention in a world increasingly shaped by algorithms.

May your preferences remain your own, and your rights are unautomated.

Acknowledgment

This work owes its shape to many minds who, in diverse ways, live the very question at the heart of this book: *nobility in an age of automation*

I am deeply grateful to the ethicists, technologists, educators, and quiet thinkers who choose depth in a world that rewards conscience. Those who feel deeply, question boldly, and act with care in environments shaped by efficiency and inner soul—you inspired this work more than you may know.

Please be advised that I tried my best, using my knowledge, ability, and available language dictionary sources to bring this message to you. So be kind, pay attention to the heart of prophecy and awareness, and do not be so particular about small possible mistakes in grammar, chosen words, and sentence structure.

Special thanks to those who challenged my assumptions, asked better questions, and reminded me that humanity is not something we encode, but something we embody.

Preface

We live in an era where machines anticipate our needs before we voice them, offer choices before we make them, and quietly shape how we see ourselves. Artificial intelligence is not merely a technological advancement—it is a force subtly redefining what it means to be human.

This book began with a simple thought: *In a world built for efficiency,* algorithms learn our habits, desires, and preferences; they begin not only to respond to them, but to form them. The danger is not that AI will conquer us, but that we may slowly surrender the very traits that make us human: agency, moral judgment, and the capacity to choose with intention.

What follows is not a rejection of AI, but a call to remain fully awake within it. It is a meditation on nobility, not as status or lineage, but as conscious choice. To be noble today is to choose wisely to act ethically and prefer what is meaningful over what is merely optimized.

Nobility is not a relic. It is a responsibility.

Contents

Chapter 1
The Essence of Nobility in Modern Times

In an age where intelligence is outsourced to algorithms and decisions are increasingly driven by data, there emerges a singular, untouchable elegance, nobility. Not the nobility of lineage or title, but of character. A quiet, commanding presence that chooses principle over profit, dignity over drama, and timeless values over trending noise.

The modern world is obsessed with efficiency. Digital assistants whisper instructions and systems predict our desires before we voice them. Yet, in this breathtaking acceleration, something essential begins to slip: the poise of the human soul.

True nobility is the new luxury. It is the final frontier that machines cannot imitate.

These are seismic acts of humanity performed beneath the surface of a world obsessed with optics. While machines calculate, the noble human calibrates, not toward profit, but toward principle.

Artificial intelligence may master logic. It may refine efficiency. But it cannot possess moral depth. It cannot cultivate elegance under pressure. It cannot choose love over indifference. That is our realm. That is our throne.

Nobility is not an aesthetic; it is a force. And in the mechanized labyrinth of the modern world, it remains our most exquisite rebellion.

Redefining Nobility Beyond Bloodlines

Once upon a time, nobility was bestowed at birth, a matter of bloodlines, banners, and the echo of one's surname through marble halls. Thrones, titles, and crests decided who would dine on silver and who would bow. But history, in its relentless elegance, has stripped nobility of its costumes and returned it to its rightful place.

Today, nobility is no longer confined to those born behind gates. It belongs to those who stand tall when the world kneels to compromise. It is no longer about what adorns your name, but what anchors your spirit. The modern noble is not clad in silk, but in silence when provoked. Not armored in chainmail, but in courage when it costs them everything.

Across civilizations, we see a common thread woven through empires and philosophies: nobility was never just a class; it was a calling. In the East, the samurai held their honor as their most sacred possession, choosing death before disgrace. In the West, true sovereign was not measured by opulence, but by justice and generosity to the powerless. Whether philosopher or prince, the noble soul walked a path where virtue reigned, and vanity dissolved.

To be noble now is to be unshakable in value while fluid in form. It is to rise above the noise, not by shouting louder, but by walking quieter, with conviction. In this era, titles fade. Influence fades. But the resonance of a noble act, however unseen, endures.

Nobility, once housed in castles, now lives wherever a human being chooses overindulgence or integrity.

The Subtle Power of Noble Acts

Nobility in the modern world does not arrive draped in ceremony. It is not escorted by title, nor does it depend on wealth, pedigree, or recognition. Instead, it manifests in the smallest of decisions, often made when no one is watching, when applause is absent, and when the right choice is the quietest one in the room.

We live in a time captivated by spectacle. Visibility has become currency, and gestures are often measured not by their sincerity but by their reach. Yet nobility, true nobility, defies this economy of attention. It moves outside the frame, unobtrusive and pure, surfacing in those brief, unrecorded instants where character either rises or recedes.

It is found in the nurse who, after twelve hours on her feet, still speaks to her patient with the dignity she herself has been denied. It is the entrepreneur who, amid rising success, refuses to compromise the integrity of their vision simply to win a quicker market. It is the activist who, confronting injustice, refuses to become cruel in the name of justice. These are not grand gestures in the traditional sense. There are no medals for these acts. No headlines. Yet they are seismic in their moral weight.

Modern nobility is made of such moments – unseen, uncelebrated, and yet infinitely consequential. It asks nothing yet gives everything: presence, restraint, and decency. It is a power that does not dominate, but anchors. It does not conquer; it uplifts.

We make mistakes under the influence of authority. But nobility teaches us that true authority is not about command –

it is about conduct. A noble person may never hold public office, may never manage a boardroom or shape public policy. Yet, the way they carry themselves – how they speak, how they choose, and how they treat others.

These people move among us. A father teaches his children to apologize not because they're weak, but because they are strong enough to face their mistakes. A barista who learns every regular's name not for tips, but out of respect. A woman who stands alone in a meeting to challenge a decision that sacrifices people for profit. These are not roles we associate with "greatness" in the traditional sense. But nobility has never obeyed convention.

Nobility lives in the texture of daily life. In the tone of one's voice. In the ability to choose patience over provocation. It is less about grand visions and more about granular discipline. It is rising to meet unfairness not with vengeance, but with values.

This is where nobility is most potent. It is not found on marble pedestals, but in unmade beds, handwritten notes, and decisions that cost us our comfort but preserve our souls. It is forged not in status, but in solitude.

Let the world race toward automation. Let algorithms predict our preferences and machines handle our tasks. But nobility—genuine, human nobility—cannot be outsourced. It belongs solely to the living. It requires choice, conscience, and courage.

To live nobly today is not to act for applause, but to live in alignment with something greater than ambition: the enduring, indelible truth that how we live, especially when no one is looking, is who we are.

And in that space, even the smallest gesture, a kind word, an honest apology, can become an act of silent magnificence.

The Internal Architecture of a Noble Life

To live nobly is not merely to act with integrity; it is to be built differently from within. Beneath gestures, beyond appearances, nobility constructs an internal world of rare symmetry: one in which clarity is more valuable than control, and dignity is non-negotiable.

This inner architecture cannot be downloaded or mimicked. It is not performative virtue or borrowed elegance. It is carved, deliberately, by choices made in silence, by restraint shown when excess is easier, and by devotion to values when the crowd drifts toward convenience. There is no fanfare in this building process. It is a private craftsmanship, more akin to sculpture than strategy. Piece by piece, action by action, the noble life reveals itself in shape and structure.

In a time when algorithms anticipate every impulse and social systems reward speed over depth, nobility returns us to slowness, to stillness, to sovereignty. Noble individuals do not react; they respond. Not because they are detached, but because they are aligned. They are not governed by chaos outside, but by an order within.

This order is not rigid. It is rhythm. It is the cadence of a person whose values are not seasonal, whose tone is not dictated by trend. Their presence is a kind of calm that interrupts the fury of modern life, not loudly, but undeniably. You feel it in the room when they speak, not at volume, but in weight. You feel it when they refuse to betray themselves, even when it would be easier to gain favor. You feel it in their eyes: unhurried, unthreatened, uninterested in winning something that would cost them themselves.

This is a strength. It is not soft, as some might imagine. It is not passivity. To choose consistency over applause, to honor principles over profit, to seek refinement over reaction, this is strength forged in silence. It is the kind of strength that remains after the storms have passed, after the voices have quieted. It is the architecture that holds when everything else collapses.

The noble life, then, is not simply moral. It is masterful. It is composed of the precision of an artisan and the conviction of a sovereign. No machine can replicate its depth. No code can imitate its soul. Because it is not assembled from data, it is shaped by devotion.

And this, perhaps, is the greatest luxury of all in our time: to live not outwardly curated but inwardly composed. To carry a palace within, built not of marble, but of meaning. And to dwell in that palace daily, with grace, with strength, with quiet magnificence.

Introducing the NOBLE

In a world where the pursuit of wealth, status, and fleeting pleasures often takes precedence, nobility is the quiet rebellion—an enduring statement that true greatness is found not in possessions, but in the way we shape our lives and the lives of others. It is not an elusive concept; it is a set of virtues that, when embraced, transforms the ordinary into the extraordinary. Here, we introduce the virtues that form the foundation of the NOBLE structure: Nurturing, Optimistic, Benevolent, and Ethical. These virtues are not abstract ideals; they are the daily decisions, the actions, and the mindset that define a life well-lived, a life of meaning.

Each virtue within the NOBLE framework serves as a pillar, a structural element that holds the noble life in place, steady and unmoved amidst the chaos. Together, they form a cohesive whole, an architecture of character that transforms both individual lives and the communities they touch. These are not virtues to be admired from a distance; they are virtues to be embodied, embraced, and enacted with courage and consistency.

Nurturing:

Elevating Others Through Care, Mentorship, and Presence

True nobility begins with the ability to elevate those around you, to recognize the potential in others, and help it flourish. Nurturing is not an act of charity, nor is it a passive gesture; it is an active, intentional force for good. It is the mentor who spends hours helping a protege find their path. It is the friend who listens not just with ears, but with a heart that seeks to understand. It is the parent who sacrifices time and comfort to ensure their child's growth.

Nurturing is not grandiose. It is in the quiet moments, the shared glance that reassures, the unexpected act of kindness, the words that heal and build confidence. It is the leader who ensures that their team not only survives but thrives. In a world driven by individual success, nurturing becomes a revolutionary act. To elevate others, to see their greatness and invest in it, is to embody the highest form of nobility.

Optimistic:

Facing Uncertainty with Hope and Creating Beauty from Adversity

In a world marked by constant change and unpredictability, optimism is not the naive belief that everything will always be easy or effortless. Instead, it is the audacity to hope in the face of uncertainty, to see beyond the current struggle and envision a future of possibility. Optimism is not about ignoring hardship; it is about embracing it with the belief that even in the most challenging times, beauty can emerge, and growth is possible.

Noble people do not succumb to despair when circumstances are bleak. Rather, they cultivate hope during adversity, not because they are immune to suffering, but because they understand the transformative power of a hopeful heart. Optimism is the quiet belief that no matter how dark the night, the dawn will always come. It is the willingness to move forward, to innovate, to dream, even when the path ahead is uncertain. In the face of global challenges, optimistic leaders inspire others to act with purpose and resilience, turning struggle into triumph.

Benevolent:

Choosing Compassion Over Cynicism, Giving Without Expectation

Benevolence is the heart of nobility, a deep and abiding compassion for others, rooted not in self-interest, but in the genuine desire to alleviate suffering and enrich lives. The benevolent person gives without expectation, without seeking recognition or reward. Their actions speak louder than words, and their generosity is not contingent on what they receive in return.

It is the philanthropist who gives not just their wealth, but their time and expertise, to uplift others. It is the friend who shows up, uninvited, in times of need. Benevolence transcends transactional relationships; it embodies a higher moral calling, a recognition that we are all connected, and our well-being is intertwined. Benevolent acts may be small, but their impact is immeasurable. In a world that too often values competition over compassion, the benevolent person reminds us that true power lies in giving, in choosing kindness even when it is not the easiest choice. Benevolence is the soul's antidote to the world's cynicism.

Ethical:

Standing Firm in Values, Even When No One Is Watching

The ethical person stands as a beacon of integrity, unwavering in their commitment to what is right—regardless of the consequences. Ethical behavior is not conditional; it is the foundation upon which all other noble virtues are built. To be ethical is to live by an inner compass that always points toward truth, fairness, and justice. It is the courage to make the right decision even when no one is watching, to honor commitments even when it's inconvenient, and to speak the truth even when silence might seem easier.

Ethics is not about perfection—it is about consistency. It is the refusal to compromise on what is right, even when the world around you encourages shortcuts, concessions, or expediency. The ethical person does not choose the easy path or the popular one; they choose the path that is just, even if it means walking it alone. They understand that nobility lies not in what is seen, but in what is done when no one else is looking.

These virtues are not relics of history or myth—they are urgently needed today. Nurturing, Optimistic, Benevolent, Leading, and Ethical: these are not distant ideals to admire but living values to embody—in every moment, every decision, and every relationship.

As we continue through this book, we'll explore how to cultivate each of these virtues in both personal and professional life. But remember: the NOBLE framework is not just a set of principles—it's a call to action. It's an invitation to live with purpose and integrity, building a life that not only uplifts you but also elevates the world around you.

Together, these five pillars form the framework for a noble life, a life that transcends mere ambition and ascends to the heights of moral greatness. They are not ideals confined to history or myth; they are the virtues that must be lived today.

Nurturing, Optimistic, Absolutism, Benevolent, Ethical—these are not qualities we admire from a distance; they are qualities we choose to embody in every moment, in every decision, and in every relationship.

As we move forward in this book, we will explore how to cultivate each of these virtues in both personal and professional life. But remember, the NOBLE framework is not merely a set of guidelines; it is a call to action, an invitation to build a life of purpose and meaning that not only elevates yourself but lifts the world around you.

Nobility as the Highest Innovation

As we stand at the precipice of a future increasingly shaped by algorithms, automation, and artificial intelligence, what will define our humanity in this new age is creativity. In a world where machines can calculate, predict, and optimize, the true disruption, the highest form of innovation, will come from within. It will come from the rebirth of nobility, reimagined through this framework.

They are not ideals confined to history or myth; they are the virtues that must be lived today. Nurturing, Optimistic, Benevolent, Leading, and Ethical—these are not qualities we admire from a distance; they are qualities we choose to embody in every moment, in every decision, and in every relationship.

Nobility, once the exclusive domain of the privileged few, is now the most radical luxury anyone can afford. It is not the luxury of wealth or access; it is the luxury of a refined spirit, an unwavering moral compass, and the audacity to live with integrity, purpose, and generosity. This is not a luxury reserved for those born into privilege or power; it is a luxury available to anyone willing to make the daily choice to act with courage, compassion, and authenticity. And in a world increasingly devoid of meaning, it may very well be the most precious luxury of all.

True greatness, as we know it, has always been defined by how we treat others, by the dignity we uphold, and by the moral courage we summon when faced with challenges. As automation reshapes industries and AI transforms the way we live and work, the need for noble human leadership, compassion, and resilience becomes more critical than ever. In a time where convenience often trumps character and efficiency eclipses empathy, nobility emerges as the antidote, an enduring force that not only elevates the individual but also elevates humanity itself.

Nobility is not an archaic ideal. It is the next evolution of human greatness, an innovation of character that will lead us into the future with purpose, clarity, and depth. As we move forward into a world where technology will continue to advance at an unprecedented pace, it is this innovation of the soul, nobility, that will define our collective destiny. It will be the force that helps us navigate the complexities of a hyper-connected, automated world with grace, dignity, and profound moral clarity.

In the chapters ahead, we will explore how to cultivate this nobility, how to build a life not only of success but of meaning, not only of achievement but of purpose. This framework, these virtues, will be your guide to navigating this evolving world, where true disruption comes not from code, but from character.

The future will desperately need nobility—and it is a luxury that everyone can afford. The question remains.

In a world increasingly shaped by algorithms, automation, and artificial intelligence, the answer lies not in the sophistication of our technology but in the elevation of our character. While machines can calculate, predict, and optimize, the true disruption—the highest form of innovation—will come from within. It will come from the rebirth of nobility, reimagined through this framework.

These five pillars form the foundation of a noble life that transcends mere ambition and rises to the heights of moral greatness.

Chapter 2
Nurturing – The Foundation of Noble Character

Nurturing is the first noble virtue, not because it is loud, but because it is foundational. It is the invisible gesture that turns raw existence into luminous becoming. It is the oxygen of all greatness—the quiet force behind every masterpiece, every leader who leads with wisdom, every citizen who stands with courage, every child who grows into a giant.

True nurturing asks for no audience. It is offering without spectacle. It is not transactional. It is sacred. It is the unseen generosity that undergoes human flourishing in a world increasingly dazzled by speed and efficiency. Nurturing is not a weakness. It is not soft. It is cultivation—the art of believing in the not-yet-visible. It is the power to kneel so another may rise.

So, it is with human character. Nurturing is not only the work of mothers and fathers. Imagine a world where it is the art of anyone, greatness was not measured by visibility, this is the economy of nurturing—not what one gathers, but what one grows.

And in this way, it builds moral beings—not with instruction, but with example. Not with pressure, but with presence. There is a quiet lineage running through the lives of noble individuals. Not all of them can name the hands that molded them. But if you trace the contours of their character, you will find evidence of early presence – someone who stayed, someone who tended, someone who knew how to see beyond the surface.

We live in an era that celebrates self-made success, but there is no such thing. Every noble soul was, at some point, nurtured. And those who become truly great often return to that posture, not just as recipients, but as givers.

To nurture is not to rescue. It is not fixed. It is to accompany, to walk alongside someone's sacred unfolding with reverence and patience. The soul does not bloom under scrutiny. It blooms under tenderness.

And so, nurturing becomes not only the foundation of noble character, but it also becomes its enduring essence. A noble person does not cease to nurture simply because they have arrived. They understand that nobility is not a throne to sit on, but a garden to keep tending.

Nurturing Beyond Bloodlines – A Collective Human Calling

Nurturing is often miscast as the private work of mothers, fathers, and familial bonds. But noble nurturing is not confined to bloodlines—it is a grander responsibility, a societal art form, a cultural lifeline. It is the invisible architecture that upholds civilizations, passed from hand to hand, heart to heart. It is a whisper through generations that says: "You are worthy of growth, and I will help you rise."

In this expanded vision, nurturing becomes a collective calling. It is not what parents alone do in the hush of nurseries, but what we all must do in boardrooms, classrooms, courtrooms, studios, clinics, and streets. It becomes a rhythm of life—a way the noble moves through the world, offering patience instead of pressure, and guidance instead of dominance.

Mentorship as Noble Generosity

True mentorship is not transaction—it is transference. The mentor does not sculpt the future; they illuminate it. Mentorship becomes the moral currency of a noble.

Educators as Modern Nobles

Consider the classroom—not just as a room, but as a realm. Within it stands a figure more powerful than any ruler: the teacher. Not the dispenser of facts, but the cultivator of futures. In underserved schools and fractured systems, educators often standalone against tides of poverty, neglect, bureaucracy, and burnout. Yet still they show up. Still, they nurture.

These are the new nobles—not by title, but by the way they wield compassion as their currency and patience as their power. They do not teach for applause. They teach transformation. And in doing so, they do not merely shape minds; they shape destinies.

And then there are the larger, more visible tapestries: global mentorship initiatives, cultural exchanges, artistic collaborations across continents. While we won't name them, we can envision them—metaphors of moral progress. Picture a global web of mentors guiding the next generation with patience, not pressure. Imagine refugees welcomed not as burdens but as potential. See communities where the elderly are not discarded but revered as walking libraries of life.

Nobility rises wherever care is chosen over convenience. Because in the end, the nobles do not merely succeed, they sow. They leave behind gardens of growth in every place their spirit touches. They are not remembered for how high they rose, but for how many others they helped to rise with them.

This is the calling. This is the charge. To nurture without limit. To care beyond kin. To live as if the growth of another were the growth of us all.

Cultivating Noble Environments in Modern Contexts

Nurturing is not a passive gift; it is an act of deliberate creation, a thoughtful architecture of space and spirit. To nurture is to design life's conditions so that growth can occur. It is an active, ongoing commitment to build environments where the seeds of potential find fertile ground and where the human spirit can stretch toward its highest expression.

It demands vision and courage to build systems that honor growth, presence, and dignity in every context.

Nurturing in the Workplace

The modern workplace is often seen as a battleground for productivity and profit, but within its walls lie profound opportunities for noble nurturing. They do not simply oversee tasks; they cultivate talent. They recognize the unseen potential in others before it blooms. They understand that nurturing a person's growth—intellectual, emotional, and moral is the most strategic investment any organization can make.

In these workplaces, success by the long-term flourishing of everyone. Here, mentoring replaces micromanagement. They know that nurturing talent requires patience, attentiveness, and an unwavering belief in the human capacity to evolve.

Education

In classrooms around the world, the true art of nurturing reveals itself in the capacity to honor each child's uniqueness. Education that embraces noble nurturing moves beyond rigid standards and standardized tests. It meets children where they are, recognizing that growth does not follow a single path or timetable.

Noble educators design learning environments that celebrate diversity of thought, pace, and style. They create spaces where curiosity is sparked, not suppressed; where questions are as valued as answers, where mistakes are embraced as steps on the path to mastery. This approach requires humility from educators—a willingness to listen deeply, to adjust expectations, and to honor the sacred individuality of every learner.

Such nurturing transforms education into a moral craft, where teachers become artisans shaping not only intellect but character. It is a pedagogy of presence and patience, where the flourishing of the whole person takes precedence over mere achievement metrics.

Moral Craftsmanship:

Designing for Growth

Each of these examples—workplaces, classrooms—reveals that nurturing is a form of moral craftsmanship. It is not a vague ideal but a tangible design challenge. To nurture is to shape environments with intention, foresight, and care, so that growth is not left to chance but becomes the natural and inevitable consequence.

This requires a mindset shift: seeing environments not as static backdrops but as dynamic forces that shape behavior and character. It calls for leaders, educators, and citizens to become intentional architects of the conditions where humanity thrives.

Nurturing is, thus, elevated from a personal act to a collective responsibility. It is the art and science of creating worlds where dignity is honored, potential is recognized, and transformation is fostered—not someday, but every day.

In cultivating such noble environments, we reclaim a sacred truth: that to nurture is to craft the future itself. It is the deliberate planting of roots that will one day bear the fruits of courage, kindness, and greatness. And in these spaces, the foundation of noble character is not merely laid—it is fortified, celebrated, and passed forward with steady hope.

Schools

When testing scores dictate priorities, the space for soul-deep engagement shrinks. True nurturing in education demands time for questions without immediate answers, for mistakes that teach, and for the slow blossoming of unique talents.

Healthcare

In the realm of healing, the tension between nurturing and efficiency reaches a poignant peak. The patient – once the center of compassionate care – can become a statistic, a brief encounter in an endless procession. The healer, burdened by time constraints and systemic demands, struggles to perform acts of deep presence and attentive listening, those essential elements that nurture body and soul.

Medicine's tremendous advances in technology and diagnostics offer unprecedented tools, yet the human touch, the pause to understand fears, to comfort anxieties, to honor suffering, faces erosion. Healing is not merely a mechanical process; it is an art requiring time, empathy, and patience. When the pace of care accelerates beyond human endurance, the risk is not only burnout among providers but also the loss of healing's deepest power.

Nurturing as a Countercultural Act

Against this backdrop, nurturing emerges as a profound act of resistance. To nurture is to reclaim time in a culture obsessed with acceleration. It is an assertion that people, human beings with complexities, stories, and rhythms, matter more than pace and productivity.

Nurturer – Noble Practices for Daily Life

It becomes clear that nurturing is neither an elusive ideal nor the sole domain of saints and sages. Instead, it is a profound virtue available to every individual willing to embrace it with intention and courage. To nurture is to commit daily acts of care, presence, and generosity—small but seismic choices that ripple outward in ways both visible and unseen. The following section invites readers to step into that noble calling, not as a distant aspiration, but as a practical, living way of being.

Nurture

In becoming a nurturer is awareness. Readers are gently guided to scan their own lives—within the intimate sphere of family, the landscape of the workplace, and the broader, vibrant fabric of community. Look around with open eyes and an open heart to discover where the quiet power of nurturing might take root.

At home, nurturing might mean choosing patience in moments of frustration, or creating a space where vulnerability is met with warmth rather than dismissal. In the workplace, it could be the deliberate act of mentoring a colleague, offering encouragement rather than competition, or fostering a culture where growth is valued over mere output. Within the community, nurturing may be as volunteering, advocating for the marginalized, or simply extending kindness to strangers whose lives brush briefly against one's own.

Reflecting on Those Who Nurtured Us

Next, readers are invited to a deeply personal exercise of gratitude—recalling those who, often quietly and without fanfare, nurtured their own growth. This may be a teacher who saw potential beyond the surface, a parent who stood steadfast in love, a friend who lent an empathetic ear. The chapter encourages writing a letter to this unseen hero, expressing thanks for the labor of care that shaped their journey. Whether the letter is sent or held privately, the act itself becomes a ritual of recognition and honor, reinforcing the interconnectedness of human flourishing.

This audit is not a tool for self-condemnation but for empowerment. It shines light on patterns and habits, inviting intentional shifts that transform everyday interactions into opportunities for moral craftsmanship. The question is not whether one can nurture perfectly, but whether one is willing to nurture consistently, learning and growing in the process.

A Lasting Legacy: Watering Others for Eternity

Let's end this chapter with a stirring reminder that nurturing is a force that transcends time and recognition. "To nurture is to leave a mark on eternity." Those who invest in the growth of others—who water tender shoots with care, patience, and unwavering faith—may remain anonymous to the world, but they are never forgotten in the tapestry of human experience. Their influence endures in the lives they touched, the characters they shaped, and the ripples they set in motion.

The virtue of nurturing stands not only as the foundation of noble character but as its fullest expression.

Chapter 3
Optimism – The Noble Vision of Possibility

Optimism, in this light, becomes courage with foresight. It is not rooted in illusion but in vision. It sees not only what is, but what could be. It is the audacity to believe in emergence, resilience, and the quiet miracle of change. To be optimistic is to reject resignation as a way of life. It means standing at the edge of uncertainty and responding with steady resolve.

This form of optimism is not impulsive. It is not reactive. It is constructed through a disciplined worldview. It is chosen. It is fed. It is guarded. It must be nourished like a fire that warms others, Noble optimism, like all virtues, is ethical in nature. It does not merely serve the self. It exists to step when others hesitate.

That is noble optimism. And it changes everything.

Cultural Catalysts –Optimism as Organizational Energy

While noble optimism begins within the individual, the most transformative expressions are found in the environments it shapes. It does not stay sealed in private conviction; it radiates, quietly but unmistakably, into the structures and systems we inhabit. When optimism moves from person to pattern, it becomes cultural energy. It fuels how teams dream, how communities collaborate, and how institutions respond to the unknown.

Optimistic leadership is not about cheerleading. It is about creating an atmosphere in which potential can breathe. Imagine a workplace where questions spark excitement rather than hesitation, where brainstorming feels less like a battle and more like a symphony. Picture a team not driven by the fear of failure, but by fascination with the unknown. In such spaces, optimism is not a mood; it is the medium in which progress is painted.

In social enterprises, optimism often takes the form of persistent creativity in the face of constraint. A limited budget does not stifle imagination; it sharpens it. These organizations do not ask, "What can we afford to do?" but rather, "What can we not afford to ignore?" Their optimism is born not of ease, but of ethical urgency. In these cultures, resourcefulness is revered. The impossible is not a wall, it is a window waiting to be found.

In tech startups and design labs, optimism pulses like electricity – sparking iteration, experimentation, and the willingness to be wrong on the way to what's right. The unknown is not an enemy, but a playground. Ideas grow faster in sunlight, and here, optimism is the sun. It encourages action before perfection. It gives permission to prototype. It asks, "What if?" instead of "What's the worst that could happen?"

In grassroots movements, optimism becomes oxygen. It sustains weary volunteers, fuels late-night organizing, and breathes life into vision long before outcomes are visible. Here, optimism is more than a strategy; it is survival. It holds communities together when systems seem stacked against them. It says, "We may not have everything, but we have each other. And that is enough to begin."

When practiced nobly, optimism spreads. It does not dominate; it inspires. Like fire catching dry kindling, it travels not by force but by invitation. A leader's belief in a better future becomes a shared possibility. One person's hope becomes a team's daring. What begins as a mindset becomes a movement.

This kind of optimism is not noisy. It doesn't always show up in slogans or mission statements. Sometimes, it is hidden in how meetings are run, how mistakes are treated, how people are seen. But over time, its presence is unmistakable. Cultures steeped in noble optimism feel different. There is levity without denial. Grit without grimness. Urgency without panic. These are not just efficient places, they are alive.

In such cultures, optimism is not about ignoring reality. It is about choosing which part of reality to multiply. Pessimism is a reaction; noble optimism is a design. It shapes rhythms, rituals, and relationships. It alters how we interpret failure and success. It turns organizational life from a competition into a collaboration – from machinery into movement.

Let us be clear: optimism cannot be imposed from the top down. It must be cultivated from the center out. It grows in leaders who listen. In teams that trust. In systems that dare to center people over process. And once rooted, it spreads. You will know it by what it produces.

Noble optimism is not a tactic. It has a texture. It is the felt sense of possibility humming beneath the surface of a system.

And when it is present, you do not need to be told; it is evident in the courage people show, the risks they take, and the joy with which they work.

It is hope, made organizational. It is a vision, made culture. It is energy, made noble.

Training the Eye –Practices for Realistic, Noble Hope

Noble optimism is not inherited – it is honed. It does not emerge fully formed in the hearts of the serene or the fortunate. It is forged in fire, cultivated through conscious discipline, and practiced most meaningfully in the valleys of uncertainty, disappointment, and fatigue. To become an optimist of noble character is not to be spared the struggle – it is to stare into it and still whisper, "There is more than this."

At its core, optimism is not simply a feeling – it is a way of seeing. Like a trained eye in a darkened room, noble optimism adjusts to perceive what others miss. It asks better questions. It scans for flickers of light others overlook. And it resists the temptation to collapse complexity into defeat.

Possibility Thinking:

The Discipline of "What Else Might Be True?"

The first practice is deceptively simple: possibility thinking. It begins with a single, powerful question in the face of closed doors, bad news, or internal despair— "What else might be true?"

This is not a denial of what is wrong. It is an inquiry into what may also be right. It asks: Amid the limitations, what remains open? Amid the grief, what growth is quietly forming? Amid the setback, what strength is being born?

Possibility thinking does not ignore hardship – it expands the frame. It dares to believe that reality is not fully captured by the worst parts of it. Like a lens shifting focus, it searches for nuance where despair offers only absolutes. It is not about spinning suffering into sweetness. It is about refusing to let pain write the whole story.

Cognitive Craftsmanship:

Holding Grief and Hope in Tension

To practice noble optimism is to become a kind of cognitive craftsman – a shaper of thought patterns that make room for both realism and redemption. This is not about "positive thinking" as avoidance. It is about building a mental framework sturdy enough to carry sorrow and possibility at once.

The craftsman does not discard painful emotions; they integrate them. They learn to say: "This is hard – and we will find a way." They honor what hurts without handing it the final word. They know that grief acknowledged becomes wisdom, and hope held alongside it becomes power.

In this light, noble optimism is not emotional suppression – it is emotional synthesis. It does not require the absence of fear, only the presence of courage. It holds the broken pieces of the world gently, and yet with hands that still build.

Solution-Seeking Conversations:

The Noble Art of Forward Vision

Another practice is the cultivation of solution-seeking speech. Noble optimism does not live in abstract ideas – it takes shape in how we talk to ourselves and others. Language becomes the tool by which futures are imagined or foreclosed.

In cultures marked by cynicism, it is easy to slip into complaint. But the noble mind refuses to stop there. It does not merely name what is wrong – it reaches for what might be made right. In conversation, noble optimism asks: "What can we do?" "What might we try?" "Who could we become?"

These are not naïve questions. They are creative ones. They are the questions that rebuild cities, reimagine policies, and restore dignity to people who were told they had none. Complaint without creation is indulgence. But critique paired with vision – that is, transformation.

To speak with hope in dark times is not to deceive. It is design.

Eyes Open, Feet Forward:

The Grounded Nature of Noble Optimism

Perhaps the most important distinction to make is this: noble optimism is not escapism. It does not pretend. It does not polish pain into platitudes. It does not rush past sorrow to arrive at sunshine.

Rather, noble optimism walks through fire with its eyes open. It knows the heat. It feels like the burn. And still it walks.

It is the parent raising a child through illness with tenderness and tenacity. It is the teacher who keeps showing up to an underfunded classroom with unwavering belief in her students' worth. It is the leader who holds fast to a better vision, not because it is guaranteed, but because it is good.

This optimism is not soft. It is not shallow. It is forged in realism and aimed toward renewal. It refuses to make hope for a product for sale. It returns hope to its rightful place – as a virtue, a discipline, a defiant act of love.

To train the eye is to train the soul. Choosing noble optimism is to build a life where hardship is not hidden, and hope is not postponed. It is to see through difficulty, not around it. And to trust that what we see – when seen rightly – can be shaped into something greater.

The Ethical Frame – Optimism as a Collective Virtue

There is a quiet temptation in modern culture to shrink optimism down to a personal commodity, to treat it as something we cultivate solely for our own well-being, our own success, our own accent. Social media slogans, vision boards, and glossy affirmations often reduce hope to a kind of motivational mirror: "If I believe hard enough, I will receive what I desire."

But noble optimism wants more. It asks more. It offers more.

It is not a trick of attraction. It is not a spiritual vending machine. It is not a cosmic contract in which belief guarantees outcome. Noble optimism is not the means to ease—it is the moral courage to pursue better.

This section reframes optimism as a collective virtue, a stance of ethical responsibility toward the future not just of the self, but of the community, the stranger, the world.

Beyond the Prosperity Gospel of Optimism

We must begin by confronting the counterfeit. The prosperity gospel, whether wrapped in religious language or disguised in secular self-help, tells us that belief guarantees blessing. That if we just stay positive, success will flow. That optimism is a magnet for wealth, health, or status.

But the truth is more complex – and more beautiful. Some of the noblest optimists in history did not gain luxury. Many walked through prisons, through exile, through failure. And yet, they clung to a hope that was not transactional, but transformational. Their optimism was not a strategy to win – it was a vow to serve.

Noble optimism does not whisper "believe and you shall receive." It declares, "Hope, and you shall endure. Act, and others may rise."

Dismantling Shallow Hope

There is also a kind of optimism that is loud but hollow – an echo chamber of endless affirmations, disconnected from responsibility. It is easy to say, "Good things are coming," while remaining passive. It is easy to chant, "Stay positive," while the world around us burns.

But optimism without engagement is escapism. And hope without action is sentimentality.

Noble optimism refuses to float above the fray. It enters the injustice, the ache, the conflict—and still refuses to surrender to despair. Not because it is blind to the pain, but because it refuses to believe that pain gets the final word.

Noble Optimism: A Stance of Service

Here is a deeper definition:

"Noble optimism is not wishing for ease—it is working for better."

It is the labor of the teacher who believes in a child that no one else sees. It is the persistence of the nurse who offers dignity to a patient that the system forgot. It is the quiet courage of the community organizer who builds bridges in a world that rewards walls.

It is not optimism for the self alone – it is optimism on behalf of others.

This is what makes noble optimism an ethical stance. It is not merely an internal disposition. It is a choice that affects how we speak, how we lead, how we raise children, how we build cultures. It is a commitment to become a living vote for the future of others. A living resistance against despair.

Planting Belief in Others

Perhaps noble optimism's highest form is this: to live in such a way that others begin to believe in their own potential because you believed first.

You become the mirror in which they see their capacity. The voice that reminds them of their value. The presence that says: "You are not alone, and this is not the end."

Optimism, then, becomes a legacy. It's something you pass on. It becomes a kind of moral contagion – hope passed from hand to hand until communities rise, and generations remember that someone once believed when it wasn't easy to believe.

That is the gift of noble optimism: it does not end with the one who holds it. It spreads. It shapes. It resurrects.

And that, perhaps, is the noblest vision of all.

Becoming a Noble Seer of Possibility

Pause.

Think back to a moment - when things were uncertain. A season of delay, disappointment, or doubt. And yet, you chose to believe in something better. You chose to act with hope. However small, that choice altered something: your path, your posture, maybe even the life of someone watching from the sidelines.

That was not sentiment. That was nobility.

Noble optimism is not about personality or positivity. It is about vision. The willingness to see what is possible – not because the evidence is overwhelming, but because the alternative is unacceptable. It is the sacred act of holding open the door of possibility long enough for others to walk through.

Let this be your quiet courage: To become a seer of potential in a world that worships predictability. To look at your community as what it could become. To see your generation as a canvas of unfinished purpose. To face the world's wounds and still whisper, "We are not done yet."

What do you still believe is possible?

Name it. Write it. Pray for it. Work for it.

Not because it is guaranteed. But because noble souls do not require guarantees to take the first step. They only need vision – and the moral imagination to believe that storms don't erase lighthouses. In fact, it is in storms that lighthouses are revealed for what they truly are: not immune, but immovable. Not protected, but purposeful.

You may not control the waves. But you can hold the light. You can be a signal of safe passage, a presence of direction, a testament of faith.

This is the noble calling of the optimist - to stand through the night with the lamp lit.

And in doing so, you become what the world most needs: a living testament to hope.

Chapter 4
Benevolence – The Heart of Noble Action

Long before the modern world debated ethics in boardrooms or taught kindness in character education curriculums, the idea of benevolence was etched into the soul of ancient civilizations. Across continents and centuries, human cultures, separated by oceans, politics, and language, arrived at a strikingly similar truth: that goodness toward others is not weakness, but wisdom; not naivety, but nobility.

Thousands of miles away, early Buddhist teachings developed the concept of *metta* – loving-kindness. But this isn't sentimental or surface-level sweetness. Metta is the disciplined cultivation of goodwill toward all beings, including enemies and strangers. In Buddhist practice, it begins as a seed planted inward— "May I be well"—and then expands in widening circles: to friends, then to the indifferent, and eventually to those who have caused harm. It is a deliberate choice to release bitterness and to recognize our shared fragility. Where chesed binds communities through duty, metta frees the soul through compassion.

Then there is *ren* – the moral core of Confucian thought. Described often as "humaneness" or "benevolent virtue," ren is the heartbeat of an orderly and ethical society. But Confucius never spoke of ren in abstract terms. He tied it to practice: to how one treats a sibling, a neighbor, a stranger. It is less about emotional warmth and more about moral excellence, the kind of inner refinement that guides outward action. In Confucian philosophy, ren is the quiet dignity of a leader who considers the needs of others before himself. It is the refined grace of someone who carries power lightly, because he knows it is meant to serve.

Three words. Three worldviews. And yet, at their center, a single radiant principle: that to live nobly is to live generously.

What is striking is not just the universality of benevolence across traditions, but its elevated place within each one. These weren't fringe virtues or footnotes in dusty doctrine. Chesed, metta, and ren were – and remain – central pillars of moral life. They represent a kind of strength that transcends personal gain: the strength to forgive, to serve, to give without keeping score.

In the modern world, where benevolence can feel endangered – flattened into slogans or mistaken for softness – this ancient chorus reminds us of something crucial: kindness is not a trend. It is a tradition. It is not a retreat from ambition, but a redirection of it. Real power, these traditions say, lies not in how many follow you, but in how many you lift.

As we explore benevolence further, we will not treat it as charity work or occasional generosity. Instead, we will reclaim it as the ethical engine of noble action, the enduring, intentional outpouring of goodwill that shapes families, cultures, and nations. Benevolence is not the extra thing we do when life slows down. It is the essential thing we must do while life rushes by.

And so we begin, not with definitions, but with a legacy – a long, golden thread of kindness that spans civilizations, reminding us that moral nobility begins not with brilliance, but with benevolence.

Kindness – How Goodness Heals the Giver

Despite all its philosophical and spiritual depth, benevolence is also ingrained in our biology. Kindness has a physiologically transformative effect in addition to being morally uplifting. Something extraordinary occurs beneath the surface of our skin when we decide to extend goodwill. Our brains light up. Our hearts settle. The body responds to generosity as if it has just received a gift, even when it is the one giving.

Science has only just started to fully comprehend what is at the core of this reaction. The first is oxytocin, sometimes referred to as the "bonding hormone." It is released in times of love, trust, and connection, such as when a parent holds a child, when a friend provides support, or when we are kind to others without anticipating anything in return. Oxytocin gives us a sense of security. It quiets fear. It tells the nervous system, "You are not alone."

The neurotransmitter of reward and anticipation, dopamine, comes next. It's what comes to mind when we solve a problem, achieve something worthwhile, or feel good. The startling thing is that dopamine increases when we do good deeds for other people as well as when we receive good things.

The effect of serotonin is less evident but no less important. Serotonin, sometimes referred to as the mood stabilizer, aids in controlling anxiety, sleep patterns, and emotional equilibrium in general. Researchers have discovered that good deeds, especially those performed consistently and intentionally, raise serotonin levels, which, over time, improve mood and mental health. It turns out that kindness is more than just an emotional vitamin. It's healing.

This is the hidden architecture of benevolence: a feedback loop between body and soul. We are better prepared to continue practicing kindness the more we do it. Furthermore, the biochemistry of benevolence increases resilience in contrast to cortisol, which rises under stress, or adrenaline, which spikes in response to danger. It nourishes rather than depletes. It strengthens without hardening.

And it doesn't end with two people. Benevolence is contagious. When someone receives kindness, they are more likely to pass it on – to a friend, a stranger, even a stranger's child. Researchers describe this as an "upward spiral" – a chain reaction of good. It begins in the nervous system, but soon it moves outward like ripples in water, shifting the emotional climate of whole groups, even cultures.

This is what makes benevolence more than a nice gesture. It is a form of social medicine. Public good. A way to heal what policy can't touch, and profit can't measure. In a world battling disconnection, distrust, and despair, acts of kindness are not distractions. They are countermeasures.

So, when we speak of benevolence as a noble virtue, we're not indulging in idealism. We're naming a force that is as chemical as it is spiritual – as grounded in our biology as it is elevated in our ideals. And every time we choose it, we become not just better people, but healthier ones. More whole. More connected. More alive.

Goodness in the Marketplace – The Economics of Care

Benevolence doesn't stop individual gestures. It can live in the bones of our systems – in the way we design, invest, and organize our collective lives. While traditional economics has long favored competition, scarcity, and self-interest as its driving forces, there's another current running quietly beneath the surface. One that's older than capitalism and more enduring than markets. One rooted in generosity, shared flourishing, and mutual responsibility.

You can see it in mutual aid, a model that dates back centuries. Long before there were formal safety nets, people formed informal ones. Neighborhoods took care of each other. Farmers shared seeds and labor. Widows were fed, children watched, illnesses covered – without paperwork, price tags, or strings attached. Today, mutual aid is re-emerging as a powerful form of economic compassion, especially in moments when institutions fall short. It's not charity. It's solidarity. A statement that says: your survival is bound up in mine.

Then there's the idea of gift economies – systems where value circulates not through profit but through relationship. In such cultures, gifts are given freely, and often anonymously, not to earn favor but to strengthen connection. This may sound quaint or utopian, but it still thrives in various communities around the world. At some spiritual retreats, for instance, teachers are not paid; they are supported through voluntary contributions. Teaching becomes a gift. The response becomes a gift. And somehow, it sustains.

Modern iterations of this spirit can be found in what some call stakeholder capitalism—a philosophy that asks: what if business existed not just to enrich shareholders, but to benefit all – employees, customers, suppliers, communities, the planet itself. It's a quiet revolution within the marketplace, and some companies are beginning to lean into it not just with branding, but with real changes in practice.

Think of an outdoor clothing company that repairs old gear for free, encourages customers to buy less, and uses its profits to fund environmental justice. Or a cooperative grocery that returns profits to local members and prioritizes local growers over distant conglomerates. These are not just branding strategies. They are decisions rooted in a different kind of bottom line.

To be clear, no system is perfect. Even the most benevolent organizations wrestle with compromise and complexity.

Benevolence in economics is not about martyrdom. It's about rethinking value – expanding the definition of success beyond shareholder returns or quarterly margins. It's about recognizing that trust, goodwill, and reciprocity aren't sentimental – they're essential. Societies that forget this become brittle. Markets that forget this lose their moral compass.

At its best, benevolence in business invites us to remember something deeply human: we are not only consumers and producers. We are neighbors. We are kind. We are stewards of the world we share.

And in that light, the workplace becomes more than a place of transaction. It becomes a place where values are lived, where generosity has scale, and were doing good is no longer a footnote – but a foundation.

The Benevolence Paradox – When Giving Gives Back

There's a strange and beautiful thing about kindness: the more you give it away, the more you seem to have. It's not a transaction. It's not a trick. But repeatedly, people who live generously often find themselves enriched – not always in material ways, but in ways that matter even more.

This is what we might call the benevolence paradox – the idea that giving, which looks on the surface like a loss, ends up multiplying something invisible but essential. It flies in the face of the old story we're told about human nature: that deep down, we're all just out for ourselves. That we only help others when there's a clear payoff. That kindness is weakness in a world ruled by scarcity.

But reality tells a different story – one that's quieter, but no less true.

Consider the feeling after helping a stranger – no recognition, no reward. Just the steady warmth that comes from knowing you did something decent. That sense of alignment with your own values. That flicker of purpose. It doesn't come from applause. It comes from within. And it lingers.

There are countless studies now showing how acts of kindness activate powerful neurochemical responses – dopamine, serotonin, oxytocin. But you don't need a lab to know what it feels like. You've lived it. We all have. A phone call made at the right time. A meal shared. These things change us, often more than they change the one we helped.

It's not about superiority. It's about connections. Something in us expands when we give. Our world is getting larger, not smaller. And paradoxically, our sense of lack begins to shrink.

This doesn't mean that every act of kindness comes with a prize. Sometimes we give and never see the fruit. Sometimes we love, and it costs us. That's real too. But even then, something in us knows it's better to live this way. It's better to care for and be hurt than to close off and calcify.

The paradox deepens when you look at people who've lost much – and somehow become more generous. Refugees who open their homes to others. Elders on fixed incomes who still give faithfully to causes they believe in. Survivors of trauma who become fierce protectors of others' peace. They don't give because they have extra. They give because they know what it means to be in need, and they refuse to look away.

All this challenges the cynical view that humans are essentially selfish. If that were true, the world would have collapsed long ago. We would have devoured each other. But we haven't. We're still here, in large part because of the quiet, consistent acts of goodness that don't make headlines but hold up the sky.

Yes, we can be greedy. Yes, we can be tribal, short-sighted, and cruel. But that's not the whole story. And maybe it never was.

Benevolence isn't some fluffy moral garnish. It's proof that another way of being is possible – and that we already carry it within us. A generosity that replenishes itself. A kind of wealth you can only measure by how much you share it.

So, here's the paradox in plain terms:

The act of giving, done freely and without expectation, often ends up giving us more than we imagine.

Not in trophies. Not in praise.

But in clarity.

In joy.

In the quiet conviction that we are not here just to survive, but to serve something greater than ourselves.

And when you live from that place, everything changes.

Living the Heart Outward

Benevolence doesn't live in theory. It lives in habits; in the quiet choices we make each day to lead with care instead of calculation.

It starts small. A warm word when it's easier to scroll past. The door held open. A meal left on a neighbor's porch without a note. The truth is that kindness doesn't need a stage. Some of its most powerful expressions are the ones no one ever sees. But they matter. They shape the air we all breathe.

At the individual level, try this: once a week, do something generous that no one asks for. Don't post it. Don't look for credit. Just let it be a gift, a reminder that the world runs better when we each take up responsibility to make it a little more humane.

In workplaces, benevolence might look like formal recognition, but it can also be as simple as asking someone how they're really doing and meaning it. It's checking in after a tough meeting. It's acknowledging effort even when outcomes are uncertain. It's knowing that culture isn't what's written on the wall – it is what people feel when they show up to work.

At the community level, benevolence scales through systems. Mutual aid groups, food co-ops, neighborhood tool libraries, these aren't flashy, but they're vital. They're the connective tissue of a society that remembers we belong to each other. In a world where institutions often fail, communities still have the power to choose generosity as a mode of survival and a model for something better.

The point is: you don't have to wait for perfect conditions. You don't need permission or a platform. Benevolence isn't bound by your income or your influence. What it asks for is attention. Intention. A willingness to orient your life toward others – not as a sacrifice, but as an offering.

When we live this way – when we build it into our days, our homes, our companies, our neighborhoods – we don't just create a kinder world. We become the kind of people who help others believe a kinder world is possible.

And that, in the end, is the real gift.

Not what you give.

But what you're giving awakens in someone else.

Chapter 5
Leading - Noble Guidance Through Example

Two Kinds of Power – A Tale of Contrasting Leadership

When the stakes are high, the situation is tense, and someone must choose the truth even though it is the most difficult course of action, that is when leadership is demonstrated. German Chancellor Angela Merkel faced a pivotal moment in 2015 as waves of refugees flooded into Europe. Politically, the safe move would have been to shut the gates. Public opinion was shaky. Tensions were high. But Merkel didn't blink. "Wir schaffen das," she said – we can do this. Her decision to welcome over a million displaced people wasn't just a policy; it was a signal. A leader had chosen compassion over calculation. And though the choice wasn't without fallout, it marked a moment of moral clarity: leadership isn't about avoiding pressure. It's about absorbing it on behalf of others.

Then there's Jacinda Ardern, the former Prime Minister of New Zealand. Her tenure wasn't perfect – no leader's is – but what set her apart was the steady presence of humanity in her governance. After the Christchurch Mosque shootings in 2019, she didn't just offer condolences—she stood among the grieving, covered her head out of respect, and spoke not of retaliation, but of unity. Her leadership was firm, kind, and deeply anchored in moral courage. Her strength wasn't performative; it was connective.

Every day. Especially when no one is watching.

First to Serve – The Quiet Power of Servant Leadership

In a world obsessed with dominance, servant leadership sounds almost subversive. And maybe that's the point.

When Robert Greenleaf first introduced the idea in the 1970s, it landed like a quiet revolution. Leadership, he argued, doesn't begin with ambition – it begins with service. Not the kind of service that hides a thirst for control under polished humility, but a genuine desire to uplift others, to listen deeply, and to build structures where people thrive, not just obey.

Take, for example, Howard Behar, the former president of Starbucks. He wasn't known for showy declarations or viral speeches. He was known for remembering baristas' names, visiting stores across continents, and asking employees what they needed to succeed. "We're not in the coffee business serving people," he once said. "We're in the people business serving coffee." That's servant leadership in action – seeing the human before the role.

Or consider Dolores Huerta, co-founder of the United Farm Workers. She wasn't seeking status; she was seeking justice. For decades, she walked with laborers, slept on floors, marched in the heat, and organized meetings no one else wanted to attend. Her leadership wasn't about the spotlight –

it was about stamina. She served first, and her influence followed.

In politics, the model holds, too. Leaders like Václav Havel in the Czech Republic didn't posture as heroes; they carried their nations through painful transitions with honesty and humility. They didn't cling to power. They bore it. For them, leadership was never about becoming more – it was about helping others become more.

What unites these examples isn't personality type or a management technique. It's a moral orientation. Servant leaders see leadership not as status, but as stewardship. They understand that power, when held well, is always borrowed – never owned.

This philosophy doesn't mean leaders become doormats or avoid difficult decisions. On the contrary, servant leadership often requires more courage. It demands that you act in the best interest of others even when it costs you. It asks you to lead with open hands, not clenched fists.

In a culture that often rewards those who shout the loudest, servant leadership whispers something more enduring: that dignity is contagious. That the best leaders don't climb over people – they lift them.

And when they do, the whole system begins to change.

The Daily Work of Noble Leadership

Noble leadership isn't about grand gestures or carefully curated optics. It's built in the small, consistent decisions that define a leader's character when no one is watching. It's how you handle disagreement, how you handle challenges, and whether your actions align with your principles when taking a shortcut would be more convenient.

First, let's talk about radical transparency. The essence of that phrase is straightforward: don't keep the truth from the people your decisions affect. It may sound like corporate jargon. Noble leaders let others in – not just when things are going well, but especially when they're not. They communicate openly about setbacks, share the "why" behind difficult choices, and resist the urge to spin. Because people can handle the truth, what they can't handle is being treated like they don't deserve it.

And perhaps most importantly, noble leadership is accountable – not in theory, but in practice. That means setting up structures where feedback isn't filtered, where critique isn't punished, and where no one is above correction. It means owning your mistakes out loud. It means answering the people, not just the boardroom or the bottom line.

These aren't soft ideals. They're hard-won disciplines. They require restraint when it is easier to dominate. They require clarity when ambiguity might protect you. They require courage, the kind that doesn't need applause.

Because in the end, noble leadership isn't about being followed. It's about being trustworthy enough to follow through.

Culture by Character – How Noble Leadership Shapes the Air People Breathe

Let's be honest: sometimes the right decision is the unpopular one. Sometimes, maintaining integrity means disappointing people you respect, slowing down progress you want, or walking away from opportunities that demand too much compromise. That's the lonely cost of noble leadership.

The legacy of a noble leader isn't written in headlines or balance sheets. It's written in the people who learned, through their example, how to carry integrity forward – especially when no one's watching.

Chapter 6
Ethical – The Noble Commitment to Principle

Ethics does not emerge fully formed. It develops with strain, repetition, and resistance, just like a muscle. Many of us start out as children with the basic notion that right is what we are told, and wrong is what we are punished for. This early moral compass orients itself around external authority—parents, teachers, rules. It's not ethics in the true sense, but obedience dressed as morality.

Psychologist Lawrence Kohlberg charted this progression in his now-famous stages of moral development. At the base level is the fear of punishment. Then comes the desire for reward or approval. With time—and sometimes hardship—we ascend into what Kohlberg called the "conventional" stage, where societal norms and laws guide our sense of right and wrong. We become good citizens, rule-abiders. But is that nobility?

Not yet.

True ethical maturity begins at the "post-conventional" level, where people are no longer simply following rules—they are questioning them. They begin to ask: *What is just? What is fair? What is true—even if the law says otherwise?* At this stage, morality becomes internalized. Principles matter more than approval. Conscience becomes louder than consequences.

In reaction to Kohlberg's theory, Carol Gilligan introduced a crucial element. She emphasized that moral reasoning encompasses care and responsibility in addition to justice and rights.

Ethics is as much a conversation as a courtroom. Kohlberg placed more emphasis on abstract principles than Gilligan did on relational ones, such as empathy, compassion, and the willingness to listen to another person's story before making a judgment.

This dual lens—principled justice and relational care—is where noble ethics takes root. Nobility is not found in blind adherence to codes, nor in the vanity of being "right." It is found in the commitment to principle when it costs something: when truth strains a friendship, when fairness disrupts convenience, when justice asks us to risk our comfort for someone else's dignity.

To be ethical in a noble sense is not to be flawless. It is to wrestle honestly with the hard questions. It is to coexist with an inner court, a tribunal based on principles rather than pride. And it is to continue to show up, day after day, decision after decision, with the silent determination to act morally even when no one cheers—and especially when no one is looking.

Many Lenses, One Goal:

Acting with Integrity

When people talk about ethics, they often imagine a fixed code — a list of dos and don'ts carved in stone. But in practice, ethical living is rarely so tidy. Life throws us into messy, tangled situations, where even our best intentions can clash. That's why thoughtful people have spent centuries developing different ways to make sense of right and wrong. Each offers a different lens, not to confuse us, but to help us slow down and see more clearly.

Some start with character. This is the heart of what's often called **virtue ethics** — the idea that who we are matters just as much as what we do. Rather than obsess over isolated decisions, it asks: What kind of person am I becoming? A leader with integrity. A friend who tells the truth. A neighbor who shows up. These aren't single acts — they're habits, shaped over time.

Then there are those who believe morality is about duty. Not duty in the abstract, but the deep inner sense that some things are simply wrong, regardless of the outcome. This is the backbone of deontological thinking — following principles not because they're convenient, but because they are rooted in respect for human dignity. Lying, manipulating, exploiting — no matter how clever the justification, these actions tear at trust, and trust is what any society stands on.

Of course, not everyone sees it that way. Some argue that what matters most is the impact. That's the idea behind consequentialism — judging actions by their results. Will this choice reduce harm? Will it make things better for the greatest number of people? There's wisdom in asking those questions. But there's also tension. What if helping the many means hurting the few?

That's where care ethics comes in — a more relational, often overlooked approach. Instead of abstract rules or sweeping calculations, it pays attention to the particular. Who is affected here? What do they need? How do we care for one another in the middle of all this? It reminds us that ethics isn't just about right answers — it's about human beings.

These frameworks don't compete. They complement. Together, they form a kind of moral toolkit. And the more tools we have, the more prepared we are — not to be perfect, but to be responsible. Thoughtful. Honest.

Because at the end of the day, noble action isn't born out of certainty. It grows from the willingness to wrestle with complexity — and still choose what's right.

Ethics in the Everyday:

Where Principles Get Personal

Talking about ethics in theory is easy. Living it? That's a different story. It comes down to the quiet choices we make day after day — the ones that don't seem like much in the moment, but they stack up fast.

Think about agreeing to an app's terms of service. Most people scroll and hit "accept" without thinking twice. But those small taps could be giving away your location, your contacts, even your habits — all so you can use something "for free." It's worth asking: Who benefits from that deal? And who might be paying the hidden cost?

Then there's the stuff we face with the planet. Do we take a reusable bag to the store just because it's the trend? Or because we get that every decision — what we eat, what we buy, how we travel — says something about what we value? No one gets it perfect. But the goal isn't perfection — it's awareness. It's asking different questions: not just "Is it cheaper?" but "Is this choice fair? Is it kind to the world?"

At work, the stakes can feel higher. Say your manager asks you to leave out a few key numbers on a report. Or maybe someone's being disrespected, and you're not sure whether to step in or stay quiet. Sometimes the challenge is subtle — like nodding along when something feels off, just to avoid conflict.

These are the moments where character shows up. It's not just about knowing right from wrong — it's about choosing to act on it, even when it's inconvenient.

Ethical living doesn't need to be loud. It could look like walking away from a deal that doesn't sit right. Or spending a little more to support a company that treats its workers well. Or choosing not to repost something that feels misleading — even if it backs your point of view.

All these choices build something. They shape not only the world around us, but the kind of person we are becoming in it.

That's what makes ethics so powerful. It's quiet. It's personal. And it's deeply human.

When It Counts:

Ethical Leadership in the Fire

Doing the right thing is simple when there is nothing at risk. When nobody is paying too much attention. When there is no chance of losing your money, your position in the room, or your status. However, the kind of leadership that counts doesn't exist in a quiet, controlled environment. It shows up in chaos. It gets tested when everything's on the line and doing the right thing isn't easy, or convenient, or popular. That's when it counts. That's when it sticks.

Take Paul Polman. When he became C.E.O. of Unilever, the goals were simple: grow revenue, expand into new markets and placate shareholders. But he would not turn it over. Boldly, he published financial results quarterly and Oriental assured concerned investors that long-term sustainability was more important than short-term profits. He seemed out of his league.

Reckless, even. But he held his ground. Under his leadership, Unilever tackled big problems—cutting down plastic waste, improving labor conditions, and pushing for real environmental responsibility. These weren't just shiny campaigns. They were backed by conviction. And despite the doubts, the company thrived. He proved that doing good and doing well weren't enemies.

Now recall the 2014 Ebola outbreak. While most of the world was either hesitant or rushed to react, Doctors Without Borders took the initiative. They didn't have to. Not because they had the advantages. They acted because they thought every life was valuable. Supplies were low. Support was late. The risks were massive. But they stayed. They worked. They faced death and chaos and did it anyway. That kind of courage isn't policy. It's not a strategy. It's something deeper—an ethical backbone that doesn't bend when things get hard. Because of them, lives were saved. Trust was built. And the world remembered what it looked like when the principal beat fear.

In tech, the lines get blurry. Innovation races ahead. Rules lag behind. And it's easy to believe that disruption, by itself, is a virtue. But people like Tristan Harris weren't convinced. A former Google design ethicist started asking hard, uncomfortable questions: What are we building? Who's really helping? Who's paying the price? He left the safety of a big tech job and spoke out about how addictive design was quietly shaping our behavior—for profit. He didn't have to do that. He could've stayed quiet, collected his paycheck, and moved on. But instead, he pushed for change. Not winning points. But because he knew silence would cost more in the long run. His voice helped shift the conversation globally.

What ties these people together isn't perfection. They made mistakes. Faced backlash. Walked uphill the whole way. But they didn't lead for applause. They were rooted in something stronger than integrity. And that gave the people around them something to believe in. Something to follow. Not because it was trendy, but because it was true.

Ethical leadership isn't about being flawless. It's about staying steady when the pressure's real. It's knowing that shortcuts leave scars. And that every decision, especially the quiet ones, teaches people what really matters. Because leadership isn't just about results. It's about values. It's about the things people carry with them long after the crisis fades or the headlines shift.

Real trust isn't built through speeches. It's built in the hard, silent moments when no one's clapping and the stakes are high. The best leaders don't just get through those moments—they define them. And when they do, their impact doesn't end with their tenure. It has become a legacy.

More Than Legal:

When Ethics Begins Where the Rules End

There's a comfortable lie many people cling to: if something is legal, it must be right. But anyone who's spent time in the real world, whether in business, politics, education, or even family life, knows how thin that assumption really is. The law sets the floor, not the ceiling. It tells us the bare minimum we must do, not the fullness of what we're capable of doing.

That's why noble living demands more than simple compliance. It demands conscience.

Think about it: slavery was once legal. So was child labor. Segregation was the law in many places. Furthermore, billion-dollar companies continue to benefit from public infrastructure, educated labor, and environmental resources while paying little to no taxes due to tax loopholes. None of those things are technically illegal. But are they right? Are they just? Are they noble?

True ethics asks more of us. It challenges us to act not just out of fear of punishment, but out of love for what's good, fair, and honorable. It's the quiet moment when someone chooses to speak up—not because they must, but because they know silence would make them complicit. It's when a company decides to protect the environment, even if regulations haven't caught up. It's when a leader tells the truth, even if it damages their brand. That's where nobility begins when we no longer need a rule to tell us what's right.

Ethical people don't hide behind loopholes. They don't ask, "What can I get away with?" Instead, they ask, "What kind of person do I want to be?" That question—simple as it sounds—is the real foundation of character. It's what separates a life of basic decency from a life of moral courage.

Following the law is necessary. But it isn't enough. To live nobly means aiming higher—not because we're trying to look good, but because we want to live in alignment with what we know to be true, even when no one's watching.

In the end, nobility isn't about perfection. It's about intention. It's about choosing the harder path when the easier one would cost us our souls. And it's about remembering that the legacy we leave will be measured not just by what we did, but by how we did, whether we rose above the minimum and chose instead to live by principle.

Chapter 7
Historical Nobility - Lessons from Aristocratic Traditions

Nobility used to mean something very different from what it does today. For centuries, it was a label stamped on a select few, the titled, the land-owning, the ones who ruled and passed down power through bloodlines rather than merit. And yet, across different cultures, the idea of nobility wasn't just about class. It carried weight. It came with expectations. Sometimes, those expectations were lived out with dignity. Other times, they self-destructed through their arrogance.

According to medieval Europe, noble families were supposed to be leaders in warfare, rulers that are just and custodians of the vulnerable. That was at least the theory. "Noblesse oblige"—the idea that the privileged owe something to the less fortunate—was more than a nice saying. It was the backbone of the entire system. Some took that seriously. They built schools, defended their people, and gave them back. Others? Not so much. They clung to power, taxed the poor into starvation, and partied while empires fell apart around them. It's no surprise revolutions came knocking.

Meanwhile, on the other side of the globe, Japan had also samurai, their very own nobility. Nevertheless, they were not wealthy lords reclining in castles. They were the warriors who believed in the rules of *bushidō*, which put value on honor, loyalty, and self-discipline. A samurai might be rich or poor, but it did not count if he lived with integrity. He was expected to put loyalty above his own life. That kind of devotion sounds noble on paper, and often it was. But sometimes, that same code demanded silence when protest was needed, or violence when compassion would've served better.

Then there's West Africa, places like Mali, the Ashanti kingdom, the Yoruba states, where nobility meant something more spiritual. Kings weren't just political leaders. Before them were custodians of culture, tradition, and moral balance. They conducted rites, mediated disputes, as well as served as a bridge between people and the gods. These rulers were selected by their lineage, of course, but they could not manage to be accepted until they had proved their value. They could be questioned/displaced when they did not serve their people. It was a sort of nobility not so in the looks so much as in the doings.

So, what do we make of all this? These systems, European, Japanese, and West African, weren't perfect. Not even close. But they weren't just about privilege for privilege's sake. At their best, they asked something more of their leaders: character, sacrifice, wisdom, courage. At their worst, they served as tools for control, wrapped in fancy titles and empty rituals.

That tension is what makes historical nobility worth studying. We don't need to go back to crowns and castles. But we *do* need to ask: what did these systems get right about holding power with integrity? What should we leave behind, and what might be worth reclaiming?

Because nobility isn't dead, it's just changed clothes. And if we want to revive the soul of it, we'll need to do better than simply inherit old ideas. We'll need to rethink what it means to lead, to serve, and to stand for something bigger than ourselves.

Lords of the Land:

Power, Protection, and the Price of Privilege

In medieval Europe, nobility was inseparable from land. To be a noble wasn't just about having a title; it was about what you owned and what that ownership obligated you to do. Land was wealth, security, and power all rolled into one. But it wasn't handed over freely. It came with strings attached.

At the heart of it all was the feudal contract. Here's how it worked: a king or higher-ranking lord would grant a chunk of land, called a fief, to a noble in exchange for loyalty and service. Not just warm feelings of respect, but actual, physical military support. If war broke out, the nobles had to show up with troops, armor, and horses. That was the deal. In return, they got to live off the land, collect taxes or rents from the peasants who farmed it, and administer justice within their domain.

In theory, it was fair trade, service for sustenance, loyalty for land. But in practice? The balance often tipped toward abuse. Some nobles took their duties seriously. They protected their vassals, maintained order, and kept roads safe. Others treated their fiefs like personal kingdoms, exploiting the people beneath them while doing the bare minimum for the crown.

Still, the structure itself was built on a network of trust and mutual responsibility. A noble had vassals under them, lesser lords or knights who were granted smaller pieces of land in return for their own service. It was a chain, with everyone theoretically bound by honor, duty, and faith.

Over time, that neat chain frayed. As kings grew more powerful and centralized authority took root, the influence of local lords began to shrink. Some tried to hold on by marrying into higher families or expanding their estates. Others simply faded into irrelevance as money, not land, became the new seat of power.

But even as the system shifted, the idea of noble obligation didn't disappear. It just changed shape. Nobility began to involve more courtly duties, bureaucratic roles, and symbolic gestures of allegiance. They weren't always riding into battle anymore, but they were expected to be examples, of taste, of culture, of governance. Whether or not they lived up to that expectation is another story.

What's worth noting, though, is that this wasn't just a system of oppression. It was also a framework, flawed, yes, that tried to tie power to service. To be granted land was to be given responsibility. And for all its problems, that's an idea we still wrestle with today: when you hold influence, who do you answer to?

Feudalism may be long gone, but the underlying question hasn't gone anywhere.

Daimyo and the Duty of Stewardship

In old Japan, long before it became a modern nation, the real power didn't sit with a central government. It lived in the hands of the daimyo, regional lords who ran their own slices of the country. People mostly remember them for their armies and castle walls. But the truth is, they did a lot more than just fight.

These daimyo weren't just sitting on top of land collecting taxes. They were expected to keep their domains running. That meant handling everything from roads to rice. If the irrigation systems broke, people starved. If education stalled, the next generation wouldn't be ready. So, a good daimyo had to think ahead. Some didn't. But the ones who cared left a real legacy.

Many daimyo used their wealth to build schools, not just for warriors but for young men training to become officials or scholars. They believed a strong domain needed brains as much as swords. Some of them brought in Confucian teachers, poets, even Zen monks. Others helped fund local temples or supported new art.

This wasn't charity. A smart, well-ordered population made a daimyo's job easier. But there was more to it than that. Cultural pride mattered. So did reputation. Supporting the arts and preserving tradition gave them status, but it also meant future generations had something to hold onto.

They also took care of the infrastructure. Roads, canals, flood barriers, those things didn't build themselves. When harvests fail, a good daimyo might open his food stores instead of hoarding them. It didn't make headlines, but it meant survival for thousands of families.

Not every daimyo lived up to that, however. Some were harsh, careless, or more concerned with power than people. But the best of them saw leadership as a kind of stewardship. They were shaping the world around them while maintaining order.

That's what made the daimyo important. Not just the wars they fought, but the lives they improved. Behind every famous name, there were everyday choices that built roads, funded schools, preserved stories, and kept communities going.

Noblesse Oblige:

When Privilege Came with a Price

There's a French phrase that stuck around long after the people who lived by it: *noblesse oblige.* It means "nobility obliges." In plain terms, if you're born into privilege, wealth, power, and title, you're supposed to carry yourself with honor. You're expected to give back, to look after those without what you were handed. Sounds noble. Sometimes, it even was.

When done right, noblesse oblige wasn't about showing off. It was about duty. A lord might fund a local hospital, step in during a famine, or risk their own safety for the people under their care. In those rare moments, being noble meant more than a family crest or a fancy title. It meant responsibility.

But let's be honest, history's full of examples where this idea went missing. There were kings who bled their people dry for wars they couldn't win. Barons who raised rents and ignored the hunger outside their gates. Noble families who clung to their status while doing none of the work that was supposed to come with it.

That's the tension built into the whole idea. *Noblesse obligation* sounds nice, but it only works if people believe in it and act on it. When they did, it built trust. When they didn't, revolutions happened.

Think of the Enlightenment thinkers or even the peasant revolts before that. A lot of those movements weren't just about wanting bread; they were about calling out the people who had the power and refused to use it for good. When nobles forgot the "oblige" part and only held on to the "noblesse," the system cracked.

But it wasn't always about grandeur or sweeping reform. Sometimes, it was small things, like a local lord stepping in to cancel a debt or defend a tenant in court. Those stories don't always turn into textbooks, but they mattered. They showed people that leadership could be rooted in care, not just control.

Of course, there were also moments when noblesse oblige was used as a mask, philanthropy for appearances, generosity with strings attached. That's still a trap in today's world, just dressed differently.

The real test of noblesse obliges, then and now, is whether the powerful are willing to do the hardest thing when no one is watching. Not just writing checks but showing up. Not just donating but listening. Not just acting for applause but standing firm when it costs them something.

In the best of hands, the idea of noblesse oblige wasn't just a social rule. It was a way of being. A reminder that nobility, real nobility, was never meant to be inherited alone. It had to be earned, day by day, in service to others.

Looking Backward to Move Forward

There's something to be said about the way some of the old noble families thought about time. They didn't just plan for the quarter or even the year; they planned for generations. They built things to last. They carried a sense of duty, not just to their own wealth or name, but to the land, the people, and the culture around them. Of course, not all of them lived up to that. Some clung to power. Some forgot the people who kept them there. But the best of them? They tried to leave something behind that mattered.

Modern leadership, for all its speed and innovation, often forgets this long view. We chase quarterly earnings, social media metrics, and rapid scaling. It's not bad to move fast, but when everything is short-term, what gets lost?

One thing the old nobles understood was stewardship. Taking care of what's been handed down and making sure it's passed on wisely. Today that could mean caring for the environment, mentoring younger leaders, protecting local traditions, or making sure a company doesn't burn out its workers for the sake of next month's report.

They also knew the value of culture. Not because it added to their portfolio, but because it gave people meaning. Many nobles funded artists, protected libraries, and supported schools, not just for their own kids, but for the broader good. Today, leaders can still do that, not by writing checks and walking away, but by showing up, listening, and making room for others to lead too.

Then there's the idea of legacy. Not as ego, but as continuity. What kind of ripples are you creating? What will your choices look like in 20 years? What kind of world are you helping to build, not just for your followers, but for their kids?

Of course, we don't want to repeat the mistakes of the past. Old hierarchies shut people out. They often concentrated power and silenced dissent. So, we take what was good, duty, care, vision, and leave the rest behind.

Being noble today doesn't mean having a title. It means taking the long view. Leading with conscience. Building something that lasts, not just for yourself, but for others too.

Chapter 8
Philosophical Foundations of Nobility

When most people think of nobility, they imagine something inherited, old families, aristocratic names, inherited land, and titles. But Plato, writing over two thousand years ago, had a different idea altogether. Plato wasn't impressed by status or family background when it came to what made someone noble. He believed nobility had more to do with how a person lived — whether they could stay honest, make wise choices, and care about justice, not just their own gain.

In *The Republic*, he described a society led not by kings or warriors, but by philosophers. These weren't people chasing titles or control — they were thinkers who had spent years learning how to question things, see clearly, and put the common good ahead of themselves. What made them fit to lead wasn't power or privilege — it was the work they'd done on themselves. For Plato, that's what true nobility looked like.

He called them "philosopher-kings"—a group selected not through inheritance but through years of education, moral testing, and personal development. They weren't born noble. They became noble.

Now, here's where it gets tricky. To support this system, Plato proposed something he called the "noble lie." It wasn't a lie in the usual sense, not a trick for selfish reasons, but more of a shared myth to hold society together. The story claimed that each person was born with a kind of metal in their soul—

gold, silver, or bronze. This wasn't about actual value, but about role. Gold souls were suited to lead, silver to help enforce justice, and bronze to build and sustain. Importantly, these metals weren't passed down through families. A farmer's child might have a gold soul. A ruler's child might not. That was the point: everyone had a place, but no one was guaranteed it by birth.

Is it uncomfortable? Definitely. The idea of using a myth to structure society sounds manipulative at best. What Plato was really wrestling with wasn't deception for its own sake — it was how to build a society where leadership didn't just fall to the rich or powerful. The so-called "noble lie" wasn't meant to trick people in a sinister way. It was more about giving the community a shared story, something to hold them together. His goal was to shift the focus away from status and onto character — to make people believe that virtue and wisdom, not birthright, should decide who leads.

And that idea – virtue over privilege – is where Plato's thinking still matters. He was one of the first to say that nobility isn't what you're born into. It's how you live. It's whether you've done the hard work of developing your mind, shaping your judgment, and learning to act in the service of others. In short, nobility is something earned.

Plato never expected his Republic to be built word for word. He knew it was idealistic. But he also knew that society needed ideals to aim for. His message, at its heart, is still relevant: if we want better leaders, we need to stop rewarding loud voices and start looking for those who lead with thoughtfulness, integrity, and the courage to make decisions for the greater good.

The Ethics of Living Well:

Aristotle's Idea of Noble Character

Aristotle didn't talk about nobility in terms of family trees or bloodlines. What interested him was how people lived — the choices they made, the habits they built, and the kind of life that slowly took shape because of it. To him, nobility wasn't inherited. It was practiced.

He believed the purpose of life wasn't pleasure or power, but *eudaimonia*, which roughly means living well in the deepest, most complete sense. It's not just about being happy now and then. It's about becoming the sort of person who, over time, feels whole. Someone who's fulfilled because they've lived with integrity.

So how do you get there? Not through luck. Not through talk. Aristotle argued that you become good by doing good again, until it becomes second nature. You want to be honest? Tell the truth, even when it's uncomfortable. You want to be generous? Give, even when no one's watching. For him, virtue was like a muscle. The more you used it, the stronger it got.

One of his key ideas was balance — or what he called the "mean." Every quality has two extremes: courage can become recklessness; humility can tip into insecurity. The noble path, he said, lives in the middle. Not lukewarm, not passive — just steady. And steady, in a world that pulls people off course, is no small thing.

He also had this phrase — *kalos k'agathos* — which means "beautiful and good." It wasn't about appearances. It was about someone whose values showed up in the way they acted, day in and day out. Someone who didn't need to announce who they were, because you could already tell.

What makes Aristotle's view still feel relevant is how practical it is. He wasn't interested in perfection. He knew people made mistakes. But he believed in growth. In trying again. In asking yourself hard questions and making small, better choices the next time.

In short, noble characters weren't a title or an accident of birth. It was something you worked on — quietly, patiently, sometimes painfully — for the whole of your life. That idea, thousands of years later, still holds power. And maybe that's because, deep down, most of us aren't looking to be perfect. We just want to be better than we were yesterday — and maybe, over time, become the kind of person we'd admire if we met them.

What Makes Someone Noble? Confucius Had a Clear Answer

In Confucius' world, being noble had nothing to do with your last name or how much land you owned. It was about how you carried yourself. Who you were when no one was watching. That's where the idea of the *junzi* comes in — often translated as "noble person" or "gentleman," but it just means someone who's decent in the deepest sense of the word.

This kind of person doesn't get there overnight. It's not just talent or luck. They put in the work repeatedly. Confucius believed that moral character is built slowly, through habits. And not the flashy ones. We're talking about the way you speak to your parents, the way you show up for friends, how you handle frustration, how you respond when someone wrongs you, and nobody else knows about it.

A *junzi* doesn't need an audience to behave well. It's not performative. They follow rituals, yes — but the rituals aren't just empty motions. They're ways of showing respect, grounding yourself, and keeping your ego in check. Think of it as everyday discipline: listening carefully instead of cutting someone off, staying calm instead of exploding, doing the honest thing even when it stings a bit.

Now, on the other side of the spectrum, there's the *xiaoren.* That's the "small person." Not small in size, but in spirit. This is the person who's mostly thinking about themselves — what they can get, how to win, how to look good even when they're not doing good. Their decisions come from self-interest, not a sense of responsibility. They go whichever way the crowd is going.

The thing is, Confucius didn't see these as fixed roles. Nobody's born a *junzi* or doomed to be a *xiaoren.* It's more like the direction you choose, again. Every situation is a chance to learn one way or the other. Confucius just wanted people to be aware of the difference — and to care about it.

It's surprisingly modern when you think about it. You probably know someone who fits both roles. The *junzi* type doesn't cut corners, tells the truth even when it's uncomfortable, shows up on time, and does what they said they'd do. The *xiaoren* will talk a big game, but when pressure hits, they vanish or throw someone else under the bus.

Confucius wasn't trying to shame anyone. He was saying: here's what integrity looks like — and it's worth working toward, even if you mess up sometimes. Because the more people who live like junzi, the stronger and more trustworthy a society becomes.

Being noble, in this sense, is less about status and more about steady self-respect. It's a quiet kind of strength. One that doesn't chase applause but earns real trust over time.

Doing the Right Thing Because It's Right — Not Because You Have To

When it comes to ethics, Immanuel Kant didn't care much about feelings or personal preferences. He wasn't interested in whether you *felt* like doing the right thing — only that you did it *because* it was the right thing. For him, nobility had less to do with emotion and more to do with intention; why your act matters just as much as what you do.

At the heart of Kant's philosophy is the idea of duty — not in a heavy, burdensome way, but as a guiding principle. He believed that every human being, simply by being capable of reason, carries a responsibility to act ethically. And he wasn't vague about what that meant. He gave us the categorical imperative — a kind of moral test. Before doing something, you ask yourself: *What if everyone did this? Would it still make sense? Would it still be good?*

For example, if you're thinking about lying to get out of trouble, Kant would tell you to pause and imagine a world where everyone lied when it was convenient. Could trust exist? Would truth mean anything at all? If the answer is no, then the act doesn't pass the test. So don't do it — not because you're afraid of getting caught, but because you respect the moral law inside you.

That's the key part: acting from duty, not just in line with duty. Let's say you help someone in need. If you only do it because someone's watching, or because it makes you feel good, that might look like a noble act — but in Kant's view, the nobility comes when you do it simply because it's *right.* Even when it's inconvenient. Even when no one sees it.

This approach is demanding. It asks for self-honesty. It doesn't give much room for shortcuts or moral wiggle room. But that's where the strength of character comes in. You're not just checking boxes or avoiding punishment. You're choosing, again, to follow a code — one that treats every person, including yourself, as worthy of dignity and respect.

There's something quietly powerful about that. It's not flashy. You won't win applause for returning the extra change a cashier gave you by accident. No one may ever know. But you will. And that consistency — that decision to live by principle even when there's nothing to gain — that's the kind of inner discipline Kant associated with true nobility.

It's not about being perfect. It's about showing up with integrity. About making decisions that, if everyone made them, would leave the world more just, more fair, more humane. That's not always easy. But nobility rarely is.

What We Carry Forward

So, where does nobility go in today's world? You won't find it in coats of arms or fancy titles anymore. But it hasn't disappeared. Some modern philosophers have tried to bring it back—not as a status, but as a way of living.

Alasdair Macintyre, for one, had a big concern: that we've lost our moral grounding. Not just in the big ways, but in the everyday stuff. We don't really know what kind of people we're trying to become anymore. He wasn't talking about laws or policies; he was talking about habits, stories, and community. The idea that we become noble not by following checklists, but by practicing certain values, over time, with other people who hold us to account. He believed that virtues need context. Without that, they kind of fall apart.

Then there's Martha Nussbaum. She took things in a different direction, focusing on real, practical dignity. What does it take for someone to live a good human life? Not just survive but thrive. She came up with a list of basic capabilities like freedom of speech, education, emotional expression, and being treated as equals. She wasn't interested in grand moral systems. She cared about whether people had the conditions to grow.

To her, nobility means taking responsibility if you've been given more. If you've got resources, knowledge, or power, the question becomes: what are you doing with it? Are you just protecting your own position? Or are you trying to make life more just for others?

So when you put Macintyre and Nussbaum side by side, something interesting happens. One says, "Find your story, your purpose." The other says, "Make sure others have the chance to live fully, too." Both care about character. Both believe that ethics isn't abstract—it's lived.

That's especially relevant now. With AI racing ahead, economies shifting, and the climate crisis making everything more fragile, we're all being asked—maybe without realizing it—to decide what matters most. Not in theory, but in the way we work, lead, vote, raise our kids, and show up for each other.

What modern virtue ethics tells us is this: character still counts. Maybe more than ever. It's not just about doing your job. It's about *how* you do it. It's about sticking to your values even when it costs you something.

In the end, nobility isn't an identity. It's a choice. Made again, in quiet ways. You don't have to be perfect. But you do have to be intentional. That's what people remember. And maybe, that's what leaves a mark.

Chapter 9
The Erosion of Nobility in Contemporary Society

If you look at the numbers from the last couple of decades, it's hard to miss what's going on. Fewer people trust the folks running the country, the news outlets reporting on it, or even the local institutions that used to feel rock solid. Big surveys — the kind done every year by places like Pew Research — keep showing the same thing: the trust line is sliding down. In some cases, it's not just a small drop. It's fallen to the point where only a handful of people still believe these institutions are working for the common good.

And it's not just about trust. Politics has turned into something closer to team sports, but without fun. People aren't just disagreeing on specific issues anymore; they're digging in, defining themselves entirely by which "side" they're on. Once that happens, talking across the divide becomes a lot harder. Even a basic compromise can start to feel like a betrayal.

Psychologists have spotted another shift, one that's harder to see but just as important. Over the years, studies of personality traits, especially with younger adults, have noticed a slow rise in self-centered attitudes. More people score higher on things like entitlement and the need to be noticed. Social media hasn't helped. They are the loud ones who get noticed now. The big, dramatic stories. The pictures where everything looks perfect. None of it really tells you much about the person behind it, but that doesn't seem to matter.

Step back and you start to see a bigger shift. What used to pass for real nobility, things like being honest, taking responsibility, putting someone else's needs ahead of your own, isn't holding up the way it used to. It hasn't disappeared completely; you can still find it in certain people and small groups. But there's less patience out there. Less trust. Less interest in anything that doesn't give an immediate payoff.

And if you want to know why, you can't just scan the headlines. You've got to look at how we live day to day, the habits we've built, the systems we work in, and the way those systems decide what gets rewarded. That's the only way to know if this drop in value is something we can slow down, or maybe even stop, before it becomes the norm.

When Value is Measured in Purchases, Not Principles

It's hard to talk about the decline of noble values without bumping into the way our economic system shapes what people care about. Consumer capitalism isn't just about selling goods anymore — it's about selling identities, aspirations, and even self-worth. Advertising doesn't simply tell you a product exists; it tells you who you'll be if you own it. The message, repeated in a thousand clever ways, is that you are what you buy.

Over time, that idea gets under the skin. It turns daily life into quiet, constant competition for status. Cars, clothes, gadgets, vacations — all of them double as markers of where you supposedly stand in the social pecking order. And if you're not keeping up, you're "falling behind, even if nothing in your actual life is worse than it was before. The yardstick shifts from moral character or contribution to community to how much you can display.

The monetization of attention has only made this worse. Every platform, app, and streaming service is fighting to keep your eyes locked in place because attention can be sold. This implies that the incentives are based on what causes the longest scroll or the fastest click, such as rage, envy, or desire. None of those promote humility, empathy, or patience. If anything, they crowd those virtues out.

This isn't to say people consciously choose greed over generosity every day. It's subtler than that. The whole environment nudges you toward constant wanting, constant comparison, and constant distraction. In a world wired that way, developing qualities like restraint, fairness, or long-term responsibility becomes an uphill climb. Nobility — the real kind, the kind that's about service and integrity — needs space to grow. And right now, most of that space is filled with noise telling us to think about ourselves first, last, and always.

How Tech Turns Conversation into a Competition

Think back to the last time a single post broke a conversation in your circle. One clip, one hot take, and suddenly, everyone's yelling in the replies. It happens fast. Too fast. And there's a reason for that: the platforms we use are built to keep us looking, tapping, reacting. They don't reward quiet thinking. They reward whatever grabs the eyeballs first.

Put plainly: attention is the point. If something makes you gasp, laugh, or rage, it spreads. If it requires thought, it often stalls. That's just how the system is wired. The software that runs your feed notices what keeps you there and then gives you more of it. Over time, people learn the easiest way to win that game: say something extreme, make it sticky, make it short. That's how outrage becomes a shortcut to visibility.

There's another trick: the game is like scoring. Hearts, shares, follower counts — little, flashing numbers that feel good to collect. People start chasing them. It changes behavior. You post what gets clicks, not what adds value. You play on the scoreboard. And the scoreboard likes heat more than depth.

The result is a culture of punchy lines and tribal applause. Nuance is messy and slow; it doesn't perform well in thirty-second clips. So, nuances get sidelined. Complicated issues get turned into slogans. Real conversation — the patient, stubborn kind that builds understanding — loses space. Instead, we get shouting matches dressed up as debates.

This also trains us emotionally. Repeated exposure to inflammatory content narrows attention and widens anger. You get used to quick moral judgments. You learn to spot enemies fast and to celebrate them being "owned." Empathizing becomes harder. The platform doesn't punish cruelty; often, it rewards spectacle.

Does that mean technology is the villain and we're helpless? Not necessarily. Tech can connect people, surface truth, and mobilize help when it matters. But the default architecture pushes toward the loudest, fastest signal. Without conscious design choices or personal discipline, the easiest route becomes the route most traveled.

So, what changes the dynamic? Small things. Slower habits. Pause before you share. Read beyond the headline. Reward people who explain instead of exploding. Build norms that applaud listening. Product designers can help too, by slowing down virality, by demoting obvious bait, by nudging toward sources that explain rather than inflame.

Until those shifts happen, expect the platform-first incentives to keep nudging public life toward outrage, spectacle, and tribalism. And expect real moral reasoning — careful, patient, forgiving — to find a way only when people decide, again and again, that it's worth the effort.

When the Ties That Bound Begin to Fray

Back then, most folks didn't work out right and wrong all by themselves. It happened in the middle of daily life with the people they were around. Family was the biggest influence, not just mom and dad, but grandparents, uncles, aunts, and cousins. Everyone had their own way of passing down what they knew. Sometimes it came as a story. Other times, it was just watching how they lived. Being part of that family web came with an unspoken push to live up to the name you carried.

For a lot of people, church or another faith community was the other steady guide. It wasn't only about rules. It was the traditions that made those ideas part of everyday life — the weekly gatherings, the yearly holidays, the big milestones. Those moments tied people together. That's where values like honesty, generosity, and kindness weren't just words. You saw them, you tried them, you loved them.

Back then, most folks didn't work out right and wrong all by themselves. It happened in the middle of daily life with the people they were around. Family was the biggest influence, not just mom and dad, but grandparents, uncles, aunts, and cousins. Everyone had their own way of passing down what they knew. Sometimes it came as a story. Other times, it was just watching how they lived. Being part of that family web came with an unspoken push to live up to the name you carried.

For a lot of people, church or another faith community was the other steady guide. It wasn't only about rules. It was the traditions that made those ideas part of everyday life — the weekly gatherings, the yearly holidays, the big milestones. Those moments tied people together. That's where values like honesty, generosity, and kindness weren't just words. You saw them, you tried them, you loved them.

Beyond family and faith, there were also the clubs, teams, and local groups that pulled neighbors into a shared purpose. It could be a volunteer crew, a sports league, or a charity drive, the kind of work that made you rely on one another. Parks, libraries, and town squares played their part too, giving people from all walks of life a place to cross paths and learn from one another without even realizing it.

But those structures have thinned out. Families scatter across cities, countries, even continents. Religious participation in many places has dropped sharply, and civic organizations struggle to attract younger members. Much of the "public square" now lives online, where ties are looser and often fleeting. This shift has brought freedoms — people can choose their communities instead of inheriting them, and old hierarchies have less grip. That can be liberating. It can also be destabilizing.

Without consistent, face-to-face connections, moral guidance becomes more abstract. Rules are debated in comment threads rather than lived out in shared work or rituals. Accountability feels weaker when your actions don't ripple through a network of people who know you personally. In that vacuum, other forces — market trends, media cycles, algorithmic nudges — start shaping values by default.

It's not as simple as saying the old ways are gone and that's bad. For some, the change has been good. They've got more freedom, more say in how they live. For others, it's left a hole. They've lost that steady hand a tight community used to give — the kind that didn't just tell you who you were but showed you every day how to live well with people around you. Now we're left asking, in this scattered kind of world, how do we build places again — real ones or online — where people can still grow, still learn what it means to live with each other?

The High Cost of Losing Our Moral Bearings

When the ties that hold people together start to fray, the damage doesn't show all at once. It creeps in quietly. A neighbor moves away, a familiar meeting place shuts down, and suddenly, the small, daily interactions that kept life grounded are gone. Many now admit they feel adrift—lonely even when surrounded by others. That sense of belonging we once took for granted is harder to find, and without it, anxiety and low spirits have room to grow. People spend more time in their own heads, less time sharing burdens or joys with others, and the result is a kind of emotional thinning.

The trouble doesn't stop at the personal level. Communities that no longer trust each other can't work well together. Political arguments turn into deep rifts, the kind that outlasts any single issue. Institutions that were once seen as the framework holding everything steady start to look like they're serving their own interests. When that happens, rules and agreements lose weight. It's not that solutions to our problems don't exist; it's that the cooperation needed to make them happen keeps breaking down.

The planet is showing the weight of our choices. Caring for it takes patience, discipline, and a willingness to act for people we may never meet. Without a shared responsibility toward future generations, problems like climate change slide down the list until they hit us with full force. The loss of guiding values isn't only something for philosophers to debate; it's a real-world problem. Restoring them isn't about longing for the way things used to be. It's about making sure we have the principles and habits we need to face the future without sacrificing what gives life its worth.

Chapter 10
Artificial Intelligence: Promise and Peril

When people bring up "artificial intelligence," a lot of folks instantly think of robots with feelings, or some machine running the world behind the scenes. Movies really push that idea, and it sticks in our heads. But day-to-day, that's not what AI is. At its core, it just means a computer taking over jobs where you'd normally expect a person to step in and decide something. Like when your phone unlocks by recognizing your face, or when you type and the next word pops up almost before you've thought it. Even those huge systems that sort through endless data to find patterns, that's AI too.

And the kind most of us deal with is called narrow AI. Nothing magical, nothing close to human thinking. It's a program built to do *one* thing well. It doesn't sit there "understanding" life or making sense of everything, it just does the job it was written for, and that's the end of it.

When Spotify gives you a playlist you like, or your GPS finds the fastest route through traffic, that's narrow AI at work. It doesn't "understand" music or traffic in the way we do. It's simply crunching numbers, looking for patterns, and predicting what will probably work.

People sometimes talk about something bigger: *general AI.* That would be a system that could learn and reason across many different areas, not just one. In theory, it would be able to adapt to new problems the way a human can—switching from solving math to understanding emotions to writing poetry. Right now, though, that's science fiction. The systems we have are clever, but they're still tools with limits.

It helps to keep this distinction in mind. If we blur the line between narrow AI and general AI, we either end up too afraid of what's not here yet, or we expect today's tools to do things they simply can't. AI isn't magic. It's built by people, shaped by data, and used in very specific contexts. The real challenge isn't whether the machines will "wake up." It's how we choose to apply what they *can* do—and whether those choices make our lives better or worse.

Real-World Benefits of AI

When people talk about artificial intelligence, the conversation often jumps straight to fears about the future. But much of its impact right now is happening quietly in areas that directly improve people's lives.

Take health care. Doctors today are working with systems that can scan medical images or bloodwork and highlight things they might miss on a first pass. The point isn't to swap the doctor for a machine, it's to give doctors an extra set of tools. An earlier warning about a shadow on a lung, or a small irregularity in a lab test, can give a patient months of time they wouldn't otherwise have. That can be the difference between manageable treatment and a much harder road.

AI is starting to show up in environmental work too. Climate researchers have always leaned on computer models, but now the newer systems can chew through way more information than before, with the help of satellite pictures, heat records, ocean measurements, all of it, much faster than people thought possible not long ago. Of course, it doesn't mean scientists can suddenly predict the future with total accuracy. What it does mean is that storm forecasts are usually sharper, and there's a clearer idea of how quickly certain places are changing. And for towns in the path of a flood or a fire, even having a couple of extra days' warning can make all the difference. People get out sooner, officials protect roads and power lines, and recovery plans start with less scrambling.

Accessibility may be the most personal area of progress. For people with disabilities, AI tools can make the world feel less closed off. When it comes to the environment, AI is already showing up in ways that matter. Climate researchers have always used computer models, but the new systems can take in way more information than before—satellite shots, ocean readings, temperature changes—all at once, and a lot faster. It's not like they can see the future with perfect accuracy, but the forecasts they put out are sharper than what we had even a few years ago.

That difference matters. If a town hears that a flood, fire, or storm might hit a few days earlier than expected, they have time to do something about it like get people out, protect buildings, set up for recovery. It's not just theory; it's the kind of notice that can save lives and money.

The Shadows of AI

It's easy to get swept up in the hype around artificial intelligence, but the problems aren't abstract. They're here, and they're already shaping daily life.

Take bias, for example. An AI system doesn't create values out of thin air. It learns from the data it's fed. If that data reflects old prejudices, the system can mirror them back. A hiring tool might start filtering out certain groups without anyone noticing. The machine isn't plotting against discrimination, but the effect is the same: some people lose out because the past was unfair.

Privacy is another pressure point. These systems thrive on personal data, and most of us give it away without much thought. With facial recognition, a person can be tracked in a crowd almost instantly. That might sound useful in some cases, but it also means everyday life can turn into a kind of surveillance state. Who controls that power? Who decides when it's too much?

Work is changing too. Automation has been around for a while, but AI moves faster and cuts deeper. Jobs built on routine decisions, answering questions, checking records, even some medical or legal tasks—are already shifting. For someone losing a paycheck, talk about "new industries" or "future opportunities" doesn't help much. The disruption feels immediate and personal.

And then there's the issue of power. A few massive tech companies own the most advanced AI tools. You never notice how it's usually just a handful of people deciding how this stuff gets used? Most of us don't even see those rooms, don't know who's in them, and yet their choices spill out into the lives of millions. Feels off, honestly, too much power piled up in a few hands.

AI isn't some disaster waiting to happen, but it's not innocent either. Acting like it's harmless is as reckless as acting like it's going to wipe us out. The only thing that makes sense is to stay awake to what's going on, keep asking the hard questions, and not let the tools end up running the people.

Living With AI in Everyday Life

For all the headlines and think pieces, most of our contact with AI isn't dramatic at all; it's subtle, and sometimes so ordinary that we forget it's even happening. Think about shopping online. You browse a couple of jackets, and suddenly the ads you see everywhere seem to know exactly what color or style you were leaning toward. That's not magic, it's recommendation software quietly shaping what you notice and maybe what you buy.

On social media, it works the same way. The posts you notice first aren't there by luck. They're chosen because the system figures they'll keep scrolling. That may not feel important in the moment, but over time it changes things. What shows up in front of us, and what doesn't, ends up shaping how we see the world, who we believe, and even the choices we make at the polls. There's no puppet master pulling strings, but there is a steady push in the background that's easy to overlook.

Education is another area that's starting to change. Students today can ask an AI to explain a math problem or outline an essay. For some, it's a lifeline like having a tutor available at any hour of the day. But it also raises questions. If a machine can draft your homework in seconds, are you really learning, or just leaning on a shortcut? Teachers now must rethink not just what they teach, but how they measure understanding in the first place.

Even relationships feel AI's touch. Dating apps, for example, rely on algorithms to suggest matches. At first glance, that's just convenience. But step back and it's worth asking: who decides what counts as a "good" match? Is it shared interests, income, looks, or something else entirely? Behind the swipe is code written by people with their own ideas of compatibility.

All of this shows something simple: AI isn't just about big scientific breakthroughs or wild stories about the future. It's already part of the everyday stuff: shopping, studying, staying in touch. Because it runs in the background, you don't always notice it, and that's why the effects can creep up on you. The real challenge isn't only making tech smarter; it's staying alert to how it's changing the pace and patterns of our daily lives.

What Makes Us Human?

One of the tougher questions about AI is whether the things we always thought only people could do are just ours. For a long time, it was easy to draw the line. Machines could crunch numbers, but they couldn't dream things up. They could follow in the steps but not feel anything. Now, though, you look around, and these systems are writing stories, making art, putting together songs, even chatting back. That old comfort line doesn't sit quite the same anymore.

Creativity is often the first battleground. If a program can generate a symphony that moves an audience to tears, is that real creativity, or just a clever remix of patterns it has learned? Many people would argue with the latter that true creativity involves a spark, a leap into the unknown that no algorithm can mimic. But others see less of a gap. From this angle, human invention also relies on rearranging what we've seen and experienced before. The question becomes: are we really that different, or are we simply more complex?

Then comes the harder subject of consciousness. We know what it feels like to be aware of ourselves, to have an inner voice, to experience joy or sorrow. Machines can't describe such inner life in any convincing way, at least not yet. Still, some researchers wonder whether consciousness might eventually emerge if a system became sophisticated enough. That possibility unsettles many, because it forces us to ask: what is the essence of being alive? Is it biology, or is it something deeper?

Moral judgment raises its own problems. We pride ourselves on being able to weigh right and wrong, but already AI is being asked to make choices with ethical weight, like how a self-driving car should react in a life-or-death situation. Should we treat those programmed decisions as "moral," or are they just rules applied without understanding? If morality is reduced to code, what does that say about the nature of our own choices?

All of these questions circle back to how we define ourselves. If machines can imitate, or even replicate, parts of what we thought made us special, it forces a reckoning. Do we double down on the belief that something ineffable separates us from the tools we build? Or do we start to see ourselves in a new light, not as beings set apart, but as part of a continuum of intelligence, natural and artificial alike?

Choosing the Path Forward

People often talk about AI as though it's going to decide our future on its own. That gives it too much credit. Technology doesn't carry intent. A hammer builds or destroys depending on the hand that swings it. The same logic applies here.

There's no doubt AI can be useful. It can help doctors spot problems faster, make transportation safer, improve how we respond to the climate, and give people with disabilities tools that make everyday life easier. These are real gains, not science fiction.

But the risks are just as real. If we're careless, the same systems can reinforce unfairness, track people without their consent, push workers out of jobs, or place too much control in the hands of a few companies. None of that comes from the code alone; it comes from the way humans choose to build and use it.

Which means the real issue is not whether AI is good or bad. The question is about us. The real question isn't whether AI will take over, but how we choose to use it. Do we handle it with care, thinking about fairness and what it means for the generations after us? Or do we rush for quick wins, brushing off the costs that come later? That's where character matters, not titles, not hype, just the willingness to act with a bit of discipline and concern for more than us.

Nobody really knows where AI is headed. It's not hiding in the shadows waiting to attack, and it's not some hero coming to rescue us either. It's a tool, nothing more. A strong one, but still, something in our hands. What happens with it depends on the choices people make. If we handle it with care and some sense of responsibility, it could end up being a real partner in our future. If we don't, it'll just carry our faults forward and make them bigger.

Chapter 11
Natural Intelligence: The Human Advantage

When people talk about intelligence these days, they usually jump right to machines. Algorithms. Neural networks. All that. But before we get carried away, it's worth remembering we've been running an advanced system ourselves for a very long time. Millions of years. Human intelligence didn't pop out of a lab. It grew out of bodies, senses, survival, mistakes, and culture piled on top of culture.

Think about it. Our brains didn't evolve in isolation. They're plugged into muscles, guts, hormones, and nerves. A stomachache can change your mood. A skipped night of sleep scrambles your judgment. You can't really separate the "thinking" from the "living." That's a huge difference from AI. A machine doesn't get hungry. It doesn't get dizzy or anxious before a big meeting. It doesn't walk into a room and notice the smell of coffee or the tension in someone's voice.

That's what people mean when they say our intelligence is embodied. We think with the whole package, not just with neurons. And it matters. If you've ever decided based on a gut feeling, you've felt that embodied part of the mind at work. Science now calls it intuition, or unconscious processing, but long before scientists named it, humans relied on it to survive. You saw tracks in the dirt, you sensed a storm in the air, you guessed which strangers could be trusted. Those weren't random hunches. They were brains and bodies learning together.

Culture is layered more on top. Language. Ritual. Storytelling. Cooking, music, tools, religion—all of that built over thousands of generations. No machine has lived through famine, war, migration, or love stories passed by the fire. We carry those lessons in habits and traditions, even if we don't notice.

Artificial intelligence, in contrast, is designed. It starts in code, sits on servers, and crunches data. It doesn't live in a body, doesn't stumble through mistakes the way a toddler does, doesn't pick up language from a parent's voice. It's disembodied. Detached. It learns by swallowing data in bulk, not by scraping a knee or watching the seasons change.

That difference shapes everything. AI can spot patterns in numbers that we'd never see. But it has no idea what those patterns *feel* like. A weather model might predict rain, but it's never been caught outside, soaked to the skin, trying to light a fire. That gap between information and experience is where human intelligence holds an edge.

We don't just process. We live, we feel, we adapt on the fly. Our cognition is messy and sometimes slow, but it's tied to survival and meaning. That mess is exactly what makes it powerful.

Many Ways of Being Smart

We often talk about "intelligence" as if it's just one thing. A score on a test. A number. But that's a narrow way to look at it. People are smart in different ways, and those ways don't always show up on paper. Some of us are great with numbers, others can't stand them but can read a room instantly. Both matters.

Take logical-mathematical intelligence. The kind that can follow rules, spot patterns, and solve equations. Sure, machines do this well. They crunch numbers faster than we ever could. But human logic isn't just cold calculation; it ties into imagination. A scientist doesn't only *solve* a formula. They *wonder* about what the formula means, what it points to. That spark isn't captured in code.

Then there's spatial intelligence. The mind's eye. The person who can picture how furniture fits in a room before moving it, or how a bridge should hold its weight. An artist who shapes clay, a pilot navigating the sky, they lean on this kind of thinking. Machines simulate space, but they don't *inhabit* it. They don't feel the resistance to wood, the drag of wind, or the thrill of a perfect curve drawn by hand.

Linguistic intelligence might seem more obvious. Words. Writers, poets, lawyers, even comedians who bend language to make meaning, or just to make us laugh. AI spits out text now, sure, but it doesn't *live* in language. It doesn't sit in silence, searching for the right word that carries memory, or pain, or love. Words are more than strings; they carry worlds inside them.

Music is another form. Rhythms in the body, melodies that can move a crowd to tears. Machines can mimic styles, yes, but they don't tap their feet to the beat. They don't get goosebumps at a crescendo. Music doesn't only note its memory, pulse, and breath.

Bodily-kinesthetic intelligence: knowing through movement. A dancer, a carpenter, an athlete. The body itself thinks through action. A basketball player doesn't calculate angles while making a shot. Their body just *knows.* Try programming that kind of knowing into silicon.

Then there are the more personal ones. Interpersonal intelligence: the ability to connect, sense, and respond to others. The friend who knows when you're upset before you even say a word. The leader who can calm a crowd. Machines can track micro-expressions or sentiment scores, but empathy? That's not in the wires.

Intrapersonal intelligence is the inward version. Knowing yourself. Your fears, your limits, your rhythms. Some people spend a lifetime learning it. Without it, all the outward smarts can go sideways.

And naturalistic intelligence: the old one, maybe the first. The ability to read the land, sense the weather, and understand plants and animals. Farmers, sailors, hunters, gardeners. People who tune into patterns bigger than themselves.

Put all this together, and you see how many layers there are. Human intelligence isn't one thing. It's a mosaic. Messy, overlapping, alive. AI is good at narrow slices, but the whole picture? The richness? That's still ours.

Thinking With Feeling

We like to pretend humans are "rational." That we think in straight lines, weigh pros and cons, and pick the best option. But that's not how it really works. Our emotions sit right in the middle of every choice we make. They're not extra. They're not decoration. They're the signal system that tells us what matters.

Fear, for example. You touch a hot stove once; you don't need a chart or algorithm to remind you. Your body remembers the pain, and that memory comes back as a jolt the next time your hand gets too close. That's intelligence. A quick, emotional warning system is faster than logic could ever be.

Joy works the same way, but on the other hand. The rush you feel when you solve a problem, or when someone laughs at your joke, it tells your brain, "Do more of that." It shapes learning, motivation, and even creativity. Machines optimize for goals, but they don't feel the kick that makes us keep going even when things are hard.

Sadness? Not useless at all. It slows us down, makes us reflect, signals that something we cared about is gone. Without it, we'd never learn the weight of loss, or the value of holding on to people while they're still here. Imagine a life without that depth. You'd have a system that moves but never really *lives.*

Then there's empathy, the glue of communities. Feeling someone else's pain, or joy, as if it's your own. That's not just "nice." It's information. It tells us who needs help, when to comfort, when to celebrate. Empathy is a survival tool just as much as fire or tools ever were.

Even anger has its place. It points to injustice, at boundaries being crossed. Left unchecked, it's dangerous, yes, but when it's channeled, it can change. Whole movements have grown out of shared outrage. Machines can detect "sentiment" in a text, but they don't burn inside with that fire.

The thing about emotions is they aren't separate from thought. They *are* thought. Or at least part of the process. They color memory, they guide priorities, they shape the meaning of every decision. Strip them away and you'll get a cold calculator, technically efficient but blind to value. That's why humans can make leaps AI can't. We *care*. And that caring bends the way we see the world.

So, when people say, "Well, AI is smarter than us in some areas," sure, it is. But it doesn't feel anything about the answer it spits out. It doesn't hesitate, regret, celebrate, or grieve. And those things aren't weaknesses. They're exactly what makes human intelligence rich enough to handle the mess of real life.

The Question of Consciousness

This is where things get slippery. Consciousness. The inner life. The voice in your head that only you hear. We can measure brain activity, map neurons firing, even predict choices seconds before you think you made them, but none of that explains what it *feels* like to be you.

Philosophers call it "qualia." The redness of red. The sharp sting of lemon on your tongue. The warm ache of nostalgia that comes out of nowhere when you smell a certain perfume. These aren't just data points. Their lived experiences. Subjective. Untouchable from the outside.

Can a machine ever have that? Right now, no. An AI can label "red," generate descriptions of sunsets, and even produce art that moves people. But it's not *seeing* red. It's not feeling the sting of memory. It's arranging symbols in ways that mimic understanding. Like an actor on stage who knows all the lines but never really inhabits the character.

Some thinkers argue that consciousness could "emerge" if systems get complex enough. The brain is, after all, just cells firing electrical signals. So why not silicon circuits instead of neurons? Maybe the difference is only material, not essential. That's one view.

Others push back hard. They say subjective experience isn't just computation. That it's tied to biology, to the body, to evolution. Maybe no matter how smart a machine gets, it will never *feel* anything. It can only simulate.

This split has been around forever in different forms. Descartes with his "I think, therefore I am." Nagel's "What is it like to be a bat?" Searle's "Chinese Room." All circling the same point: intelligence is not the same as awareness. Knowing *about* pain isn't the same as hurting. Knowing how to describe joy isn't the same as laughing until your stomach aches.

It matters because it shapes how we treat these systems. If an AI writes a poem, do we see it as creative? Or just a clever trick? If one day a robot says, "I'm afraid of being turned off," do we believe it? Or do we chalk it up to programming? At what point, if ever, do we owe machines moral consideration?

Right now, the safer bet is this: machines don't have inner lives. They process. They predict. They echo. But they don't wake up at 3 a.m. with a knot on their chest. They don't fall in love. They don't wonder why they exist.

Maybe someday that will change. Or maybe it never can. Either way, consciousness remains the real frontier. Until we can answer what it means to *experience* life from the inside, the gap between humans and machines stays wide open.

Holding on to the Human Edge

Here's the thing: we're not machines. We're animals who've been grinding it out on this planet for millions of years. That matters. Our intelligence wasn't designed in a lab; it was carved by survival. Every sense, every instinct, every shortcut in our brain exists because it worked, because it helped us stay alive long enough to pass something on. AI doesn't have that history. It doesn't bleed. It doesn't get hungry. It doesn't sit up all night worrying about a sick child. Those things shape the way we think, even when we don't notice.

Take moral reasoning. People like to imagine ethics as some logical game—you weigh outcomes, compare options, pick the least harmful. But in real life, it's messier. You feel guilty. Or pride. Or loyalty. Those emotions steer decisions in ways math alone can't. Sure, machines can run equations faster, but will they ever understand why someone chooses kindness when cruelty is easier? Or why does a mother sacrifice herself for her child? That kind of judgment grows from being human, not from solving for X.

Creativity is another one. A program can remix patterns, spit out a new tune, paint something that looks like Van Gogh. Impressive, yeah. But real invention often comes from being *stuck*. From frustration. From not knowing what to do and then stumbling into something wild. Humans get bored, restless, even desperate and out of that mess comes breakthroughs. Fire. Stories. Jazz. The stuff that changes how we live. You don't code boredom into a machine. You don't simulate desperation. Those are ours.

And then there's culture. We adapt. Not just as individuals, but together. We tell stories, pass along traditions, and learn from each other's mistakes. A tool might store all the data in the world, but it doesn't sit around a table, argue late into the night, laugh at old jokes, or cry over a shared loss. Culture isn't just information. It's a connection. And connection shapes intelligence in ways raw computation can't touch.

So yes, AI will get sharper. No question. It'll do things we never imagined. But it won't erase what makes people matter. Our biology, our emotions, our clumsy, creative way of figuring things out—that's not going anywhere. If anything, the rise of machines makes those traits even more valuable. Because in a future filled with algorithms and automation, the most human parts of us, the moral compass, the spark of imagination, the ability to grow and adapt together—those will be the edge we keep.

Not perfect, not clean. But human. And that's the point.

Chapter 12
The Nobility of Human Emotions

Science has been poking around in the brain for a while now, trying to figure out why we care about each other at all. One of the discoveries that caught people's attention was mirror neurons. These little cells fire not just when you do something, but when you see someone else do it. Watch a person stub their toe, and your brain lights up as if it happened to you. It's not imagination. It's wiring.

That wiring explains why we wince when we see someone get hurt, or why we feel a kind of lift when we watch someone succeed. Empathy isn't just a nice word. It's physical. Your body carries the echo of what you observe. This makes compassion more than a choice. It's built in. You can resist it, numb it, even twist it, but the basic machinery is there.

There's more to it than mirror neurons. Whole networks in the brain tie into this capacity. Areas that process emotion, memory, and even decision-making all pull together when we connect with another person's experience. That's why empathy can feel heavy. You're not just noticing someone's suffering, you're partially living it. It drains you, but it also links you in a way that's hard to fake.

Machines don't have this. They can be programmed to recognize patterns of sadness in a voice or spot tears on a face. But that's recognition, not resonance. A program doesn't get a knot in its stomach when it hears bad news. It doesn't lie awake because it can't stop thinking about a friend in pain. That gap matters.

And here's the kicker. Without empathy, morality gets shaky. Rules are one thing, but rules without feeling are brittle. When you care about someone, decisions shift. You go out of your way. Your sacrifice. You forgive. Empathy gives rules a heart, and without it, justice turns cold.

It's easy to forget this in a world that prizes speed and efficiency. Compassion slows you down. It demands time and energy. But it also makes life livable. It's the reason we don't just walk past suffering as if it's a glitch in the system. We feel it. We carry it. And in carrying it, we hold on to what makes us human.

That's nobility, not in some lofty sense, but in the raw biological fact that our brains are tuned to care. Machines can calculate. We can feel. That difference, small as it sounds, is enormous. It shapes everything.

The Weight and Power of Our Feelings

Emotions don't just sit inside us. They spill out. They connect, divide, heal, and sometimes break things apart. Forgiveness, grief, hope, outrage. They're not just feelings. They shape how people live together.

Forgiveness sounds simple when you say the word. In practice, it's hard. Someone hurts you. Maybe they say sorry, maybe they don't. You decide whether to keep carrying the weight or to put it down. Families only last because someone chooses to forgive. Same with friendships. Even countries.

Think of truth and reconciliation after years of violence. It isn't perfect, but without forgiveness, anger never loosens. It festers.

Grief is different. Heavy. It comes after loss, and there's no skipping it. People gather at funerals, memorials, and gravesides. They cry, share stories, and sit in silence. Grief brings people close in ways nothing else can. It teaches us that life has value. That absence leaves a mark. Without shared grief, communities turn shallow, like nothing really matters.

Then there's hope. Strange thing, really. Invisible, but strong. It keeps people moving when logic says stop. A worker in a rough job, an activist in a dangerous place, parents trying to build a better life for their kids. They lean on hope. Not because everything is fine, but because they need to believe it could be. Without it, despair wins. And despair kills faster than hunger.

Moral anger has its role too. Outrage at injustice. Without it, nothing changes. People see cruelty, lies, exploitation, and something inside says no. That can be loud enough to shake systems. Marches in the street. Rewritten Laws. Power challenged. Of course, outrage can burn out or get twisted, but when it's real, it's a force for repair.

These emotions don't run separately. They overlap, build on each other, and create meaning. They keep us human. They tell us when to let go, when to hold on, when to fight, and when to keep going. Machines can fake the signals of emotion, but they don't feel it. They don't carry grief, they don't forgive, they don't hope. And without that, they don't shape a moral world.

For us, emotions aren't extra. They're part of how we think. How we survive. How we stay together. They may hurt, but they're also the reason we build anything worth keeping.

Where Our Feelings Come From

Emotions didn't just pop up in people out of nowhere. They have roots. Old ones. If you look at other animals, especially social ones, you see hints of the same patterns. A wolf will guard its injured packmate. An elephant will linger over the bones of its dead. Monkeys protest when food is given unfairly. These aren't accidents. They're survival tools, baked into behavior.

Humans just took those early patterns and turned the volume way up.

Take care of others. In mammals, mothers already nurse and protect their young. That instinct spreads across groups too. Primates groom each other and watch over each other's kids. Now look at us. We raise children for nearly two decades before they're fully independent. Without deep, lasting care, we wouldn't make it. A human baby left alone doesn't last long. We survive because people stick around, even when it's hard. That constant caring became part of who we are.

Fairness is another one. Chimps have been shown to get angry when they see another chimp get a better reward for the same job. They'll even throw the food back. It looks like a tantrum, but it's really a sense of justice. In humans, that feeling became more complicated. We write laws, create systems of rights, and argue endlessly over what's fair. But the seed of it, that gut reaction against unfair treatment, is ancient.

And cooperation. Maybe the most powerful piece of all. Other animals hunt together and defend territory together. But humans pushed cooperation to a level nothing else has touched. We build cities, organize armies, and invent governments. We tell stories that glue thousands or millions of strangers into a single group. None of that works without emotions steering it. Trust, loyalty, shame, pride, those feelings hold groups together when logic alone wouldn't be enough.

Evolution didn't give us emotions as decoration. They were tools. A species that cared, that punished cheaters, that worked in groups, had a better shot at survival. Over time, the emotions got stronger, more layered, more tied into culture. Songs, rituals, myths—all ways of amplifying feelings so bigger groups could stay connected.

What makes humans different isn't that we have emotions and animals don't. It's that we built whole worlds out of them. Civilizations, traditions, moral codes. Our feelings became the backbone of culture itself.

Strip those away, and people fall apart. That's why machines, however advanced, don't really compare. They can calculate, mimic, and predict. But they don't care. And caring is what kept us alive in the first place.

Real vs. Fake Feeling

Pretending to feel and feeling are two different things. People usually notice the gap. You can tell when it's real, like a friend who breaks down at a funeral because they truly cared. And you can tell when it isn't, like a politician tearing up on stage just to look human. The body language, the tone, it doesn't carry the same weight. We're wired to sense the difference.

The noble part of emotion comes from it being authentic. You can't fake compassion. Not really. People might smile, nod along, whatever—but they don't buy it. The real thing shows up on its own. Little cracks. A slip in the voice. A look in the eyes. That's where you see empathy. Joy. Grief. The stuff you can't stage. A pause. Eyes that can't quite hold steady. Words that stumble instead of sounding neat. That rough edge is what makes it real.

Now look at the world we've built around social media. Platforms reward display, not depth. Post the right emoji, hashtag the right cause, and you look like you care. But it's thin. It's fast. It's a flash of feeling, often recycled from someone else's script. And the more people do it, the harder it gets to separate the genuine from the stage.

Virtue signaling is the term that gets thrown around, sometimes unfairly, but it captures something real. It's when the goal is not emotion itself, but the performance of it. "Look at me, I'm on the right side." That's not the same as sitting with someone who's hurting or changing your own life in response to a belief. The noble weight of emotion is lost when the point is image.

But to be fair, technology didn't invent this. People have always expressed emotions. Courtly manners, public rituals, staged apologies—they've been around forever. What's different now is the scale and speed. Billions of eyes watching at once. Algorithms push outrage or sympathy depending on what gets clicks. It amplifies the fake at the expense of the real.

And yet, real emotion still cuts through. You can feel it when someone shares a story that cost them something, when their words shake, when they don't polish the edges. That kind of honesty still resonates, even in a flood of curated noise. Maybe more so, because it's rare.

Authenticity isn't a perfect expression. It's messy, inconsistent, and sometimes even makes us look bad. But that's the point. Being noble in emotion means risking vulnerability, showing the truth of what we feel, even if it isn't popular or beautiful. Machines can simulate the words or the tears, but they can't make them mean anything.

Practicing Emotional Nobility

Understanding your feelings isn't enough. Just noticing them doesn't automatically make you wise or make your actions noble. You must work at it. And that work isn't fancy. It's small, slow, ordinary, and repeated every day. But it changes the way you act. It changes the way you show up.

Mindfulness is a start. Sitting quietly, paying attention to your breath, feeling your body, noticing emotions as they rise and fall. Don't try to push them away, don't try to hold them. Just see them. It sounds simple, but it isn't. Thoughts jump in, old memories flare up, impatience creeps in. That's the practice. You learn to notice without losing yourself. Over time, you start responding rather than reacting. You start choosing. That's where emotion meets action.

Perspective-taking is another tool. Not just imagining what someone else might feel but stepping into their world for a few moments. What pressures do they carry? What hopes? What small joys shape their day? You don't need to agree, or like it, or even fix it. You just try to see it clearly. It softens anger, tempered judgment, and opens space for patience. Like any muscle, it strengthens the more you use it.

Learning the words for what you feel matters too. Not just "happy" or "sad," but frustrated, anxious, wistful, hopeful, resentful, curious. Naming an emotion gives it shape. Once you can name it, you can move through it. You can act rather than react. You can respond with intention.

Finally, there's honest expression. Let yourself be seen in ways that fit the moment. A quiet word, a note of thanks, a conversation admitting you don't have all the answers, a shared laugh, even a tear. The world is full of polished, performative signals. Don't just copy that. Be messy. Be human. That's where nobility lives.

These practices aren't about being perfect. They are about expansion. Expanding awareness, empathy, patience, courage, and connection. They are small acts repeated over time that make a difference. That's what noble character asks of us: to feel, to reflect, to act—and to keep trying, every day.

There's a piece of this we rarely talk about. Emotions don't just stay inside us. They are moving. They spill over into the people around us. One person being patient, honest, or kind can change a room in ways you don't notice right away. Little things. Tiny gestures. A nod, a word, a look that says, *I see you.* Over weeks, months, years, those small actions pile up. Suddenly, the place feels different. People trust a bit more. They help a bit more. They expect a bit more from each other. That's why learning to handle your feelings matters. Not just for yourself. For everyone you touch.

It isn't simple. Old habits are stubborn. Anger flares. Fear hits fast. Envy, pride, shame—they creep in without warning. Sitting quietly, paying attention to yourself, doesn't make them vanish. Awareness doesn't mean you won't stumble. But it does give me a pause. A chance to say: Alright, I see it. Now what do I do? That pause, brief as it is, can be the difference between snapping and speaking, between holding a grudge and forgiving, between shutting people out and leaning closer.

And it's messy. Emotions aren't neat. Hope fights with fear. Joy mingles with frustration. Love jostles with sadness. You can't sort them cleanly. It's like trying to untangle a knot that keeps twisting as you pull. That's exactly why it matters. The nobility of feeling isn't in perfect control. It's in being present. Feeling the storm inside, making choices anyway, keeping your heart open even when it hurts.

It shows up in tiny daily things. A check-in call. A kind word to someone struggling. Saying no to gossip. They seem small. Almost invisible. But they build the same thread that runs through everything noble in emotion. It doesn't need applause. It doesn't get headlines. But over time, it shifts how you move through the world.

And here's the truth. Machines can't touch this. They can mimic. They can pattern-match. They can even copy our gestures or words. But they can't feel the knot in your stomach when a friend suffers. They can't know the quiet lift of hope. They can't weigh grief against joy in the messy, personal way we do. That's ours. That's the human edge. Feeling, acting, connecting, that is what makes us more than code. The nobility of emotion isn't optional. It's the backbone of life that matters.

Chapter 13
Careers and Skills in the AI Era

Let's be honest. The job markets are moving faster than anyone wants to admit. It feels like every month, something shifts. Automation isn't some far-off idea — it's already creeping in. You don't really notice it at first. Just a new update here, a change in how things get done there. Then someone's job quietly disappears. Look around — offices, factory floors, cash registers, even hospitals. Tasks that once needed people are now run by code. It doesn't make a scene. It just happens. One day it's normal.

If you dig into the numbers, it's not random. Jobs built on repetition are the easiest targets. Data entry. Routine accounting. Basic translation. Simple legal reviews. Anything that runs on clear patterns or predictable outcomes. Machines eat that for breakfast. They don't get bored. They don't need breaks. They just keep refining faster each year. The pattern is clear in every study you read. The lower the creativity and human judgment involved, the higher the automation risk.

But that's only part of the story. There's another side that often gets buried under the panic headlines. Certain kinds of work still depend on something machines can't touch. Decision-making under uncertainty. Reading the room. Knowing when someone isn't saying what they really mean. A nurse adjusting tone to calm a frightened patient. A teacher noticing confusion in a student's eyes. A negotiator who senses hesitation and shifts their pitch mid-sentence. You can't code that.

When economists talk about "non-routine cognitive tasks," this is what they mean. Work that blends thinking with emotion, strategy with empathy. It's the human mix that doesn't fit neatly into an algorithm. These roles don't vanish easily because they rest on understanding, not just execution. That's why creative fields, leadership, and social work still stand strong. They bend with change instead of breaking.

The gap keeps getting bigger. You can see it everywhere. In factories, a single robot now covers what used to take a whole team. Customer support lines? Half the time, you're talking to software before anyone really picks up. Even in newsrooms, short reports and sports updates are being churned out by scripts, not people. It's strange to watch. A little scary, too. But maybe that's the point. The issue isn't stopping it — that ship's gone. The real question is how we adapt.

The edge now belongs to people who can mix reason with imagination. Tech with empathy. The ones who can speak both machine and human. Machines are learning to mimic us, but they still need guidance, vision, and purpose. The next generation of careers won't just be about keeping up with AI. It'll be about staying irreplaceably human.

When you strip things down, what really keeps humans relevant isn't how fast we calculate or how much data we can handle. It's the things we do that machines can't quite pin down. The things that don't fit into neat boxes. Six of them, really, stand out when you look closely at the future of work.

The first is creative innovation. Not the kind where you just rearrange old ideas, but the spark that comes from chaos and curiosity. The kind that shows up in an artist sketching something no one has seen before, or an engineer solving a problem that doesn't have a clear formula. AI can remix what exists, but it can't feel the frustration of staring at a blank page or the rush of a sudden breakthrough. That tension is where invention lives.

Then there's complex problem-solving. Real life doesn't hand out clean data sets. It's full of missing pieces, conflicting goals, and unpredictable people. Figuring things out when the answer isn't obvious takes more than processing power. It takes intuition, risk, and the willingness to be wrong for a while before getting it right. Machines aren't built for that kind of messy persistence. Humans are.

The third is emotional intelligence. It sounds soft, but it might be the hardest skill of all. It's the ability to listen between the lines, to sense what someone feels before they say it. You can't fake that with algorithms. Real trust is built on tone, timing, and presence. A person who can steady a tense room or help others feel seen will always have an edge, no matter how advanced the tech becomes.

Fourth is ethical reasoning. Every major decision comes with tradeoffs, and AI can't weigh values—it just optimizes outcomes. We're the ones who ask, should we—not just can we. That question changes everything. The stuff we build, the choices we make, the lines we decide not to cross... that's what shapes the world that comes next.

And it's never just one field anymore. Climate, medicine, tech, and teaching all bleed together. You can't fix one without touching the others. Real progress shows up when somebody starts linking weird things together, like physics and emotions, or numbers and art. The people who mess around in the spaces between fields, who don't stay in one lane, they're usually the ones who move stuff forward. And finally, embodied skill. The things you can't do without a body. A surgeon's touch. A carpenter's sense of pressure. A dancer's control of breath and space. These are forms of intelligence that come from living in a body that feels and reacts. Machines can't replicate that.

These six abilities, rough, human, irreplaceable, are what keep the lights of our species burning bright in an automated age.

Jobs That Stay Human

Some jobs will bend and survive. Others will break. The ones that stay are the ones that lean hard on what only humans can do. You can already see the pattern forming.

Take healthcare. Machines can scan, diagnose, even predict, but they can't hold someone's hand while they get bad news. A nurse adjusting a pillow, a doctor calming a frightened patient, a therapist sitting through someone's silence. That presence, that touch, it can't be coded. Hospitals and clinics might fill up with smart tools, but they'll still need people who bring warmth into cold rooms.

Teaching is another one. Real education isn't just dumping information. It's a connection. A good teacher reads the room.

They notice when a student's confidence cracks, when someone's pretending to understand. They adjust. They inspire. That spark of belief doesn't come from a screen. You can run lessons through software all day, but only one person can make a kid care. Mentors, coaches, tutors, that kind of work isn't going anywhere. If anything, it's needed more now.

And the creative side? Writers, designers, musicians, filmmakers. Yeah, AI can copy the look, the sound, even the rhythm. But originality hits differently. It comes from lived experience, from loss, from humor, from odd little human quirks that machines can't fake. People don't connect with perfect creations. They connect with the imperfect ones that show truth. The cracks make it real.

Counselors, social workers, mediators—all rely on emotional depth. They listen, interpret, and guide. Algorithms can suggest solutions, but they can't earn trust. And trust is everything in that kind of work. The same goes for people who deal with others when things get rough, pastors, teachers, social workers, anyone really. That mix of calmness and care isn't something you learn from a book. You either have it, or you don't.

Even in tech, the human part still counts. Always does. The engineer who bridges teams. The project leads those who keep things from falling apart. The person who understands not just how systems work, but how people inside them do.

And the hands-on ones, builders, makers, craftspeople, those jobs stay human. It always will. There's something sacred in creating with your hands. Machines can reproduce but not create meaning. A potter's thumbprint on a bowl, a chef seasoning food by instinct, a mechanic listening to an engine and knowing what's wrong without a scan. That's embodied intelligence.

These careers, scattered across fields, share one thing: they need presence. Mind, heart, and body working together. That's the thread running through the future of work. Not efficient. Humanity.

When the Technical Becomes Human

The technical world isn't disappearing. It's changing shape. The jobs won't vanish overnight, but the kind of work people do inside them will feel different. Engineers, coders, data scientists, they're all being pushed to think less about machines and more about people.

Take engineering. It used to be mostly about solving performance and cost. Get it built, make it strong, make it work. But now the question is shifting. How does this affect lives? Who gets left out? Ethical design isn't just a nice phrase; it's becoming part of the blueprint. Engineers are starting to ask tougher questions about the systems they create. Should they? Not just can they. That's new territory.

Programmers, too. Once, it was all about logic and speed. Now they're realizing the code they write decides what people see, how they communicate, and what they believe. A user interface can calm or frustrate. An algorithm can comfort or manipulate. That's power. It needs awareness, not just technical skill. The best coders of the next decade will be part psychologist, part artist, part ethicist. They'll write with empathy, not just precision.

And then there's data science. The field exploded because data felt like the truth. But raw numbers don't tell full stories. People do. Now data scientists are being asked to translate. To take all those models and graphs and make them human. What does this mean for housing policy? For privacy? For fairness? The real challenge is not finding patterns—it's deciding what to do with them responsibly. Turning insight into wisdom instead of control.

Some of this feels uncomfortable. Technical workers weren't trained for moral judgment. They were trained to optimize, to debug, to make things efficient. But efficiency has limits. A perfectly efficient system that forgets the human element becomes cold, even dangerous. That's why so many companies are hiring ethicists, sociologists, and philosophers to sit beside engineers. It's not fluff. It's survival.

The future of technical fields will be collaboration. Teams that blend hard logic with soft insight. The machines will handle the grunt work—calculations, predictions, simulations. The people will handle the meaning. The "why." That part can't be automated.

So, no, technology won't erase engineers or coders or analysts. It'll ask more from them. More reflection. More responsibility. More imagination. The ones who thrive won't just be good at what they build. They'll be thoughtful about why they build it, and who it truly serves. That's the new skill set, human awareness inside the code.

Staying Ready for What Comes Next

Nobody has this figured out. Every time someone says they do, the world flips again. A new tool. A new system. Suddenly, the plan doesn't fit anymore. So, the best thing you can do is stay awake. Keep your eyes open. Learn what you can, even when it doesn't seem useful. Read weird stuff. Ask people how they do their jobs. Pay attention to what actually works.

Learning is noticing. It's trying something, messing it up, trying again. Most people stop doing that once they get comfortable. Don't. Build a mix of skills. One solid thing that feeds you, a few side things that keep you flexible. You don't have to be great at all of them. Just enough to move when things shift. A teacher who learns code. A mechanic who studies business. A nurse who writes. Small layers.

And people. Never forget people. Real connections. Not followers or contacts. Folks who know your name and remember how you treated them. They're the ones who'll call you when everything goes sideways. You can't fake that. You earn it. Adaptability sounds fancy but it's just patience with chaos. Something breaks; you breathe. You look around. You figure out what's next. That's all. Machines don't panic, but they also don't care. You still can, and that's a strength.

Every few months, check yourself. What feels alive? What feels dead? Maybe it's time to learn something new or move somewhere different. No one will tell you when to start. You just start. Don't waste time trying to predict the future. You won't. The trick is to stay human inside it. Curious. Kind. Steady when everyone else freezes. That's how you last. Not by outsmarting the machines, but by being the one thing they can't be. Alive, awake, willing to keep learning when everything else is changing.

Chapter 14
Noble Leadership in Business and Politics

People like to talk about ethics as if it's some nice extra, like sprinkles on top of real business. It's not. It's the thing holding the whole structure up. Strip it away, and everything starts to rot from the inside.

There's data on this. Lots of it. Companies that live by their values do better. Not overnight, but overtime. Studies from Harvard, Deloitte, and others keep saying the same thing. Firms with strong ethical cultures have lower turnover. Their people stay longer because they trust the place. Customers stick around too. They feel seen, respected. You don't need to buy loyalty when you earn it.

Ethical doesn't mean timid. It means grounded. It means you make choices that hold up under pressure, not just in quarterly reports. Look at the long haulers. Patagonia. Unilever. Companies that think in decades, not months. Their people know what they stand for, and it shows.

Now compare that with the quick hitters. The ones that cut corners, fake results, or treat workers like parts. They might have a good run, but it doesn't last. Eventually, something breaks. Trust, reputation, the team itself. Once that's gone, it's hard to rebuild.

You can feel it in a workplace too. In an ethical company, there's less fear. People speak up. Ideas move around. There's energy. In the other kind, everyone's walking on eggshells. Decisions get slower. Creativity dries up. It's like the air gets heavy.

What's interesting is that the numbers match the feeling. Ethical companies outperform their peers financially over the long term. Productivity, profit margins, customer satisfaction. All higher. Not magic. Just cause and effect. People who feel respected work harder, stay longer, care more.

It's the same in politics, though messier. Leaders who act with some basic integrity might not grab headlines every week, but they build stability. Communities trust them. That trust becomes cooperation, and cooperation builds progress. When leaders lose that moral compass, the damage spreads far beyond one term or one office.

In the end, ethical leadership isn't idealism. It's survival. The ones who play the long game always win out. It might take years, but truth compounds just like interest. The hard part is staying steady when shortcuts look easy. But that's what separates the real leaders from the ones just filling a seat.

Leadership That Lifts, Not Commands

When you think about leadership, most people picture authority. Someone at the top giving directions, setting deadlines, and keeping everyone in line. But the kind of leadership that changes people feels very different. It doesn't shout. It doesn't threaten. It draws you in.

Real leaders move people by example, not pressure. You can sense when they mean what they say. There's no hidden angle, no trick. Real leaders don't make their vision about themselves. They make space for everyone in it, so people feel their work counts — even the small parts no one notices. That's what keeps a team moving when things get tough.

The ones worth following think past the moment. They doubt their own ideas, listen before they speak, and give others room to be wrong without fear.

When people feel trusted, they stop hiding. They start trying. Ideas begin to flow, and problems stop being dead ends.

What stands out most, though, is how these leaders treat people. They notice things others overlook. A tired look in someone's eyes. A bit of hesitation in a voice. They pull that person aside, check in, and make sure they're okay. It's not written in their job description, but it makes all the difference. That kind of care turns workplaces into communities.

Integrity holds it all together. It's not the kind of leadership that needs attention. It's the kind that shows up when no one's watching. A good leader admits when they're wrong, says sorry, and keeps their word even when it costs them. People can sense that steadiness. It makes them want to do better because they trust what they see.

Fear might get people moving, but it never keeps them close. Respect does. Trust does. A title can make someone listen, but only for a while. What lasts is belief — the kind that grows when someone helps you see your own strength.

Real leaders don't see followers. They see people standing beside them. They see partners. They build others up until they can lead too. And when that happens, when a team begins to lead itself with honesty and courage, that's when leadership becomes something larger than one person. It becomes culture. And that, more than any title or strategy, is what truly lasts.

The Everyday Work of Noble Leadership

You can tell what kind of leader someone is by how they handle truth. Most people duck it when things get messy. A few don't. They face it, even when it costs them.

Being honest isn't neat. It's saying, "Here's what's working. Here's what's not." No spin, no big speech.

When leaders do that, everyone breathes easier. People stop guessing. They start trusting. That's how loyalty shows quietly, over time.

Then there's how decisions get made. Some leaders talk about teamwork but still make every call behind closed doors. Noble leadership looks different. It's slower, yes, but stronger. It means pulling people in. Listening, even when opinions clash. The process can feel uncomfortable. It takes patience to sit through disagreements, to let others poke holes in your ideas. But when people see that their voices shape the outcome, they start to care more deeply about the result. That's what real inclusion looks like.

Accountability is the hardest test of all. Anyone can talk about integrity when things go well. The real measure is what happens when they don't. A noble leader doesn't look for someone to blame. They say this part was on me. I missed something. Let's fix it. That one act of courage changes the tone of an entire group. It gives everyone permission to own their part without fear. The team becomes honest because the leader is.

And then there's credit. It's strange how rare generosity can be at the top. Some leaders hoard praise like it's money. The noble ones do the opposite. They pass it around. They call out the quiet worker who solved a small but crucial problem. They make sure others are seen. Loyalty doesn't grow from pressure. It grows when people feel seen and respected. When they know they matter, they give more of themselves — not out of fear, but belief. Building that kind of trust takes time. It's quiet work, the kind that never makes a report or slide deck.

But in the end, that's what holds everything together: trust, not fear.

Good leaders aren't perfect. They just stay honest when it'd be easier to hide. Lifting others higher even when nobody's watching. That's the quiet power that keeps a culture alive.

When Leadership Shapes the Air, Everyone Breathes

You can tell when a workplace feels safe. It's not about posters on the wall or speeches in meetings. It's the air. People speak up. They laugh. They admit when something's off without flinching. That's what real psychological safety looks like. It doesn't happen because someone writes a memo. It grows when leaders show they can take criticism without snapping, when they listen without pretending to know it all. Over time, people stop guarding every word. They share ideas that aren't polished yet. That's where creativity starts. Not in perfection, but in the freedom to sound a little wrong before you get it right.

Noble leadership makes that kind of space. It doesn't crush curiosity. It protects it. And when people feel safe to try, they also feel safe to warn. That's where ethical whistleblowing comes in. Nobody wants to be the person who raises a hand and risks their job. It takes guts. But when the tone from the top is honest and fair, when people know their concerns won't be buried or punished, they speak up. That's how corruption dies early, before it spreads. It's not about policies as much as it's about tone. A leader's reaction in those moments decides whether truth becomes part of the culture or disappears into fear.

Purpose ties all of it together. Money keeps the lights on, sure, but it doesn't pull people out of bed in the morning. Purpose does. The kind that feels personal. Leaders who act with nobility help others see why their work matters. A nurse who saves lives. A teacher shaping minds. A programmer building tools that make life easier, safer, and more connected. When people can trace a line from their work to a larger good, they give more of themselves. They take pride, not just in results, but in meaning.

The shift doesn't happen overnight. It builds slowly, conversation by conversation, decision by decision. A leader listens. Admits. Encourages. Protects. Soon enough, the habits catch on. Coworkers start to mirror the same honesty, the same courage. That's how culture really changes. Not by force, but by example.

Noble leadership turns organizations from systems into communities. It replaces fear with trust, silence with dialogue, and survival with shared purpose. It's not flashy, but it lasts. And in the end, it's what separates a place that burns people out from one that helps them grow.

When Integrity Is Tested

You only know what kind of leader someone is when things start falling apart. Smooth sailing hides a lot of flaws. It's the storm that shows who's real. Take the case of a mid-sized manufacturing company that nearly went under a few years back. Sales dropped. People were scared. The new CEO didn't sweep in with big slogans or a shiny turnaround plan. She started by meeting everyone, floor by floor. No cameras, no staged talk. Just questions. Listening. Within weeks, she found the real issue wasn't the product. It was trust. Managers had been hiding bad numbers. Workers felt disposable. Instead of firing her way through it, she opened the books. Every figure. Every cost. Radical transparency. People braced for layoffs, but something strange happened. Once they saw the truth, they started helping fix it. Ideas came from the floor. Efficiency rose. Within a year, the company wasn't just surviving. It was steady again.

Politics offers tougher ground. During a regional crisis a few years back, a governor faced both protests and financial collapse. Most in his position would've doubled down. Blame someone else. Deny everything. He didn't. He faced the crowd and said, "We messed up, and I take the blame." The room quieted. For the first time, people listened. Later, he brought in teachers, local workers, and community voices to rebuild. It wasn't perfect or fast, but it worked—because it came from honesty, not image. It was built on shared ownership.

Then there's community leadership, the kind that doesn't make headlines. A small-town split in two over a factory expansion. Half the residents wanted jobs. The other half feared pollution. Meetings turned out ugly. One local teacher stepped in, not as a mediator, but as a neighbor. She brought both sides into the same gym one night and asked them to start with stories instead of arguments. Families talked about asthma. Others about unemployment. Slowly, faces softened. They didn't agree, but they began to hear each other again. Eventually, a compromise was formed. Smaller plant. Better filters. Local oversight. No one got everything, but everyone got heard.

Real leadership doesn't always come from the ones with titles.

It often starts quietly—with someone at the bottom who refuses to ignore what's wrong.

A junior analyst who speaks up when everyone else stays silent.

A nurse who pushes back when shortcuts risk a patient's safety.

Those moments don't make headlines, but they matter. Small choices that spread further than anyone expects.

In one logistics firm I visited years ago, a young manager refused to fake delivery reports to meet deadlines. Her boss told her to "adjust the numbers." She said no. It nearly cost her the job. But that stand changed something. The next week, a few others followed her lead. Within months, the culture shifted. People began telling the truth about missed deadlines instead of hiding them. The company didn't crumble. It improved. Customers noticed the honesty. The numbers got better because trust replaced fear.

That's how it often happens. Not through big speeches, but through one person deciding to stay honest in a room that rewards pretending. You can't build an ethical culture with slogans. You build it through small acts of courage that catch on.

I've seen it in politics too, though it's harder there. A young staffer leaks the real budget numbers because hiding them would hurt public schools. A mayor refuses donations from a contractor known for shady deals. It's never clean or easy. They lose friends. Sometimes they lose their jobs. But when they walk out, they can still look people in the eye. That counts for more than people think.

The truth is that noble leadership isn't glamorous. It's exhausting. It means more sleepless nights, more hard talk, and fewer shortcuts. It means being patient when anger is easier, calm when everyone else wants to scream. And still showing up.

What sets these people apart isn't perfect. It's that they keep showing up even when it feels pointless. They don't always win. But they make space for something better to grow. That's the quiet legacy no one puts on a résumé — the kind that outlasts any title or contract.

And when enough people lead this way, something shifts. The tone of the place changes. You start hearing "we" more than "I." People step in to help instead of waiting to be told. Meetings feel less like battlegrounds and more like shared work. It's not a miracle. It's practice. Repeated, flawed, human practice.

That's what noble leadership really is. It's not some perfect virtue. It's persistence in doing the right thing even when no one claps. The mistakes still come, but the heart stays steady.

That's what noble leadership looks like when the pressure is on. It's not smooth. It's not perfect. It's human. It stands its ground without losing its soul. And when it works, it leaves behind something sturdier than victory. It leaves trust.

Chapter 15
Education for Nobility

The way we teach has drifted off course. Somewhere along the line, education stopped being about forming people and started being about producing credentials. Test scores, rankings, degrees. All the noise that looks like progress but feels empty once you scratch beneath it. Walk through most schools now, and you can sense it. Kids memorize facts they'll forget by summer. Teachers under pressure to hit targets. Everyone chasing numbers that say "success," but no one asking what kind of human being we're actually building.

It's not that tests are evil. They just became the whole story. Students learn how to perform instead of how to think. They chase grades instead of meaning. When everything gets reduced to points and averages, curiosity fades. Fear takes over. Fear of failing, of standing out, of thinking differently. That's the quiet damage no one measures.

I've seen young people crumble because they missed a decimal on a math test, but never once get asked how they'd handle a moral crossroad. We teach them to calculate, not to choose. To compete, not to care. And we wonder later why workplaces feel cold, why leaders cut corners, and why empathy feels rare. The truth is that those habits start early. You can't expect moral courage from people who've spent their whole lives being rewarded for compliance.

There's a deeper loss too. Schools used to shape character. Not perfectly, but intentionally. There were teachers who talked about right and wrong, not as punishments but as choices that define a life. Now that kind of talk makes people uncomfortable. Too personal, too messy, too "unscientific." So, we stick to data and pretend it's enough. It isn't. Numbers can't raise a generation with heart.

When education forgets the soul, everything else falls flat. A child might grow into a brilliant engineer or a skilled doctor, but without moral grounding, that brilliance can bend the wrong way. We've seen it happen. It happens in stances that smart people build harmful systems because no one ever taught them to ask whether they should.

What if schools cared as much about conscience as competence? What if the report card had a space for integrity, for kindness, for courage? Not as grades, but as questions worth asking. How did you help today? What did you stand up for? Who did you listen to? That's how you raise thinkers who can also feel.

Real education should build more than resumes. It should build humans who can hold steady in a world that keeps testing their values. Until we get back to that, we'll keep turning out experts who know everything except how to live well.

There's more than one way to teach someone how to live well. Every culture has wrestled with the same question: how do we form good character, not just sharp minds? The answers may look different on the surface, but underneath they're remarkably similar. Each begins with the idea that wisdom isn't something poured in from the outside — it's drawn out from within.

Take the old Socratic way. It wasn't about lectures or memorized rules. It was built on questions — one after another. Socrates never offered ready-made answers. Instead, he turned people back on themselves, pressing them to question what they thought they knew. It was rarely pleasant work — digging into your own reasoning never is — but that discomfort was the lesson. It stripped away arrogance and replaced it with humility, the kind that comes when you finally admit how little you truly understand. And from there, genuine insight begins.

Confucian teaching feels quieter, more inward. It's about self-cultivation—building virtue step by step, through small acts repeated daily. A respectful bow. A kind word. A promise kept. The idea is that character grows from habits, not speech. You don't need an audience. You just live rightly, again and again, until it becomes who you are. In that way, Confucian ethics make goodness feel ordinary, not heroic.

Then there's Indigenous storytelling. Whole worlds carried in words. Lessons aren't told straight; they're wrapped in story, in seasons, in the rhythm of community. The listener must lean in, to hear the layers. A child might listen to a tale about a trickster or a hunter, and years later realize what it really meant. Storytelling doesn't force understanding—it lets it ripen. That's the kind of patience the modern world could use more of.

And now we have social-emotional learning programs—our modern attempt to rebuild what older systems once carried naturally. They teach empathy, self-awareness, and resilience. Useful things. But sometimes they get boxed into lesson plans and bullet points. The heart of it, though, is still ancient. Learn yourself. Respect others. Act with awareness. The form changes, but the goal stays the same.

When you look across these traditions, a pattern shows up. Each one says character isn't taught like math. It's lived. It takes time, reflection, and connection. You learn from questions, from elders, from mistakes. You learn by watching, by doing, by trying again. Real education for nobility doesn't shout. It shapes. It invites people to become better without telling them they have to. Different paths, same truth. The measure of learning isn't what you know, it's who you become.

Rebuilding Education from the Inside Out

If schools want to shape better people, not just better workers, the way we teach has to shift. Grades and certificates aren't enough. Students need a reason to care — a real link between learning and life. The simplest place to start is how we structure it.

Project-based learning makes the classroom part of the real world. Instead of practicing on hypotheticals, students work on problems that genuinely exist. Maybe they will design a plan to reduce food waste in their town, or interview local elders about how communities used to work together. When learning has a human cost or benefit, ethics stop being abstract. A student has to think, *What's the right thing to do here? Who's affected by my choice?* That's when moral reasoning starts to grow.

Peer teaching does something different. It puts responsibility on students to help others understand. It's one thing to know something; it's another to guide someone else through it. The act itself teaches patience, humility, and empathy. You start seeing how people think differently, where they get stuck, and what kind of encouragement actually helps. A good peer teacher learns as much as they give. That habit of helping others becomes a kind of quiet leadership later in life.

Community service, when it's done right, ties knowledge to humanity. Not as a punishment or a box to check, but as a way of learning who we are in relation to others. Working in a food pantry, tutoring younger kids, planting trees, caring for elders; all of it carries lessons books can't touch. But reflection is key. Students need space to think about what those experiences mean. Otherwise, it's just another task. Teachers can ask simple questions: *What surprised you? What changed your mind? What would you do differently next time?* That's where empathy takes root.

And we can't forget the arts. Music, painting, theater, writing, they open doors inside people that logic alone never finds. Art forces honesty. You can't fake a real performance or a line that feels true. Through it, students learn to see from another's point of view, to express what can't be measured in numbers. Arts keep us human.

Education that builds nobility isn't about turning everyone into a saint. It's about balance. Heart and mind working together. Students learning that knowledge means little without wisdom. If schools can nurture that connection, we might finally raise a generation that knows how to think clearly and feel deeply at the same time.

Teaching Wisdom in a Digital World

The internet isn't just a tool anymore. It's the air young people breathe. They grow up surrounded by screens, scrolling before they can even spell. The problem isn't the technology itself. It's how unprepared we've been to live inside it. Schools teach coding and computer skills, but not how to think clearly in a digital storm. Kids learn how to search but not how to see what's real.

Digital literacy should mean more than checking facts. It's learning how to *pause.* To ask, "Who made this? Why?" before clicking share. The web is full of noise, half-truths, and clever lies dressed up like news. Without training, even smart minds get pulled in. Algorithms are designed to keep us hooked. Outrage spreads faster than truth. It's not an accident. Someone profits from every click. Teaching students gives them power—the power to choose what they feed their attention to.

Critical thinking needs to be taught like a muscle. Not once, but daily. Have students compare headlines from different sources. Let them trace how a story mutates online. Show them how data can be twisted just enough to sound convincing. The goal isn't to make them cynical, but to awake. To build an instinct for honesty in a world built to distract.

But awareness isn't enough. Ethical reflection has to go hand in hand with digital skills. Ask them: *If you could go viral, would you still post it if it hurt someone? What's the cost of being anonymous online?* Let them wrestle with the questions instead of feeding them answers. Real growth happens in that discomfort.

Technology also needs to be used to build character, not just to consume attention. Digital storytelling, online volunteering, and learning communities— these can connect people across oceans and give meaning to connection. A student helping another with math through a shared forum learns compassion in real time. A class creating digital art about climate change learns both creativity and responsibility. It's not the screen that defines the outcome. It's the intention behind it.

If schools can teach students to see technology as a reflection of themselves—a mirror of their choices—they'll start using it differently. Less about ego, more about purpose. Less about reaction, more about response.

The internet won't slow down. It'll only get louder. But if we teach the next generation to listen carefully, to think before they share, and to use their voice with care, maybe the noise won't win. Maybe, just maybe, they'll turn it into something that builds instead of breaks.

Learning That Never Ends

Education isn't supposed to stop when school ends. Learning was never meant to stop. It's supposed to spill into everything we touch. Somewhere along the line, people forgot that. We turned it into a checklist. Finish this, get that paper, move on. The moment you call yourself finished, you start to shrink. The world keeps moving. You have to keep up.

Books help. Not the ones everyone posts about, but the kind that messes with your head a little. The ones that make you stop mid-sentence and stare at the wall. Old thinkers. Strange poets. History that hurts reading. Even slow stories that take pages before they mean anything. They change the way you see things.

Reading slows you down in a good way. You sit still. You listen. You argue with a voice that isn't yours. Sometimes you hate it. Sometimes it hits too close. Either way, you come out different. That's what matters.

Then there's mentorship. Not the formal kind with contracts and titles, but the humankind. One person who's walked a little farther down the road, helping another find their footing. You learn things from mentors you can't get from any book. How to keep your dignity when you fail. How to stand firm without turning hard. How to forgive. Good mentors don't give answers. They share scars. And if you're lucky, you'll get to be that person for someone else one day.

Contemplative practice matters too. Quiet time. Space to think without noise. It doesn't have to be meditation on a mountain. It can be a walk before sunrise. A few minutes breathing before the day begins. Writing down what you're grateful for or what you regret. The goal isn't perfection. It's awareness. Most people move through life reacting. Slowing down lets you respond instead. It turns learning inward, makes it real.

And then there's community. That's where everything you learn gets tested. It's easy to sound wise alone. Harder when you're surrounded by people who disagree, who challenge you, who need you. Real growth happens in service. Join in something that pulls you out of yourself. Volunteer. Teach. Help organize a local project. When you do, you start to see how learning isn't personal, it's relational. You grow because others push you to.

Lifelong learning isn't about stacking skills or chasing relevance. It's about becoming a fuller human being. The person who listens more carefully. Who sees complexity instead of taking sides. Who keeps trying to understand instead of pretending to know. That's what education was meant for all along—to make us humble and wise at the same time.

The truth is that the world doesn't reward this kind of learning much. It doesn't fit neatly into resumes or LinkedIn updates. But it changes how you move through everything else. You stop chasing certainty and start looking for connection. You stop measuring worth in wins and start finding it in meaning.

Maybe that's the final lesson. Staying noble isn't about never failing. It's about being teachable. Staying curious even when life gets heavy. Asking questions long after the easy answers fade. If we can hold on to that—reading, listening, reflecting, showing up for others, then education never really ends. It just deepens. And that's what keeps the spirit alive, no matter how old we get.

Chapter 16
Digital Nobility: Ethics in Technology

The online world has twisted how we act, talk, and even feel. It moves too fast. Everything's instant. You tap a screen, something happens miles away, and that's the end of it. No eyes looking back. No voice to soften the edge. Nothing to remind you there's a real person on the other side. That space, that silence, it dulls something inside. It makes it easy to forget there's a heartbeat out there. Things we'd never say face-to-face, we throw out like it's nothing. A joke. A comment. A job. And suddenly it spreads, carried by anger that moves quicker than thought.

It's not like this problem just appeared, but it's heavier now. Social media, automation, all this new technology stretched our reach but shrunk our sense of care. We can touch millions but barely feel the touch back. One post can mess up someone's life. One sloppy line of code can shut down hospitals or leak data. Even a quiet choice, something small in a design, can decide who's seen and who's ignored.

Technology doesn't just grow. It multiplies. Every tool carries the best and worst of us at once. It can lift someone up or tear them apart. The same network that spreads kindness can also start a wildfire of hate. And most of us don't even stop to think about it. We just keep scrolling, keep posting, keep building. It gives us power without the usual friction that reminds us to think twice.

The scale of it is what makes it hard to manage. In older times, mistakes were local. You could apologize. Fix them. Now errors go global before you even realize what happened. A glitch in a financial system, a privacy breach, a deepfake—it can spiral in minutes. You can't pull it back. You can't explain to everyone who it reached. That scale creates a kind of moral numbness. When harm feels too big to grasp, we stop feeling it.

Then there's the problem of opacity. The systems running our lives aren't built for understanding. Algorithms make choices we can't see, can't question. They filter jobs, news, justice, and healthcare. The logic hides behind trade secrets or technical fog. People like to shrug and say, *That's just how the system works as* if it showed up on its own. But it didn't. Systems don't build themselves. People do. And every person who touches them leaves something behind. A value. A bias. A blind spot. You can see it in the code if you know where to look. Most folks don't. They just use it. And when the decisions hide inside the machine, the people behind them disappear too. That's when things start to slip. No one feels responsible anymore.

We love blaming the tech. It's easy. It feels cleaner that way. But the truth is, this isn't a machine problem. It's us. We made tools that move faster than our own sense of right and wrong can keep up with. Every update, every version, it speeds up again. It's like we forgot to build in a pause button.

Nobody stops long enough to ask, *Should we be doing this? Should we even want what we're chasing?* There's always pressure to go faster, to make it cheaper, smarter, more efficient. That word 'efficient' people toss it around like it's a good thing by itself. But efficient at what? At replacing thought? At cutting out the messiness that makes us human. Sometimes it feels like we're racing to nowhere, polishing the machine while the soul rusts underneath. But nobility isn't speed. It's a reflection. It's asking who might get hurt before we act.

If the old moral compass came from looking someone in the eye, the new one must come from imagination. We have to learn to feel the ripple of our digital actions even when we can't see it. To picture the faces behind the data. To remember that responsibility doesn't vanish in code. It just hides there, waiting for someone willing to look.

Building Ethics into the Machine

There's a growing push now to bring ethics back into the room where technology gets made. Not as a slogan, but as part of the design itself. Some call it value-sensitive design. It sounds fancy, but it just means asking, who gets helped and who gets hurt by what we build? And then actually changing the design based on that answer. It's slow work. Messy too. You must talk to real people, not just run tests. You listen. You adjust. You realize your first idea wasn't as harmless as you thought. That's what progress should look like.

Then there's this idea of algorithmic justice. It's one of those phrases that floats around at conferences, but it matters. Algorithms decide so much now. Who gets a loan. Who gets seen in a search result. Who the police stop. Bias doesn't disappear inside code—it hides there. So, people are trying to make those systems transparent, explainable, and fair. Not perfect, but less blind. The hard part is getting companies to admit that fairness costs something. Time. Money. Maybe a little pride.

Digital well-being metrics are newer. They ask questions that sound simple but cut deep. Does this app make people feel better or worse about themselves? Are they more connected or more alone after using it? How long before healthy use turns into addiction? These aren't things you can measure easily, but that's the point. Once you start counting what really matters, it changes what you make.

The challenge is all these frameworks—good as they are—don't mean much unless people in power use them. Ethics can't just sit on a slide deck. It has to live in the process, meetings, and deadlines. The coder who speaks up when something feels wrong. The manager who listens instead of brushing it off. The company that decides "good enough" isn't good enough anymore.

We built this digital world with our hands. We can rebuild it with our conscience too. But only if we keep asking the hard questions, even when everyone else wants to move faster.

Choosing What to Build

It's one thing to talk about ethics in technology. It's another thing to live it when a paycheck depends on saying yes. Most people working in tech want to do good, but the system rewards speed and profit. That pressure makes silence easier. Yet noble practice starts in those small, private moments when you choose not to look away.

Refusing to build something harmful sounds heroic, but often it's quiet. You walk out of a meeting and decide you won't code a feature that manipulates users into staying online longer. You stop pretending data collection is harmless. You speak up when a product hurts the people it claims to help. Maybe you get ignored. Maybe you don't. Still, you said something. That's where integrity begins.

Accessibility should be non-negotiable, but it's still treated like an afterthought. True inclusion means building for people who don't fit the default. The colorblind user. The person without stable internet. Someone who navigates with voice or touch instead of sight. Designing for them isn't charity. It's intelligent. When you make technology easier for those at the edges, you make it better for everyone.

Transparency sounds simple too. Just tell the truth about what your system can and can't do. But that honesty can make investors nervous. It slows momentum. Yet hiding limitations always catch up later. The trust you build by being upfront is worth more than any early hype. Users remember who told them the truth.

Then there's the hardest part: making humans flourish above profit. Every company says it wants to help people but watch where the budget goes. Does it fund the safety team or the growth team? Does it fix the harm or just the optics? A noble technologist keeps people at the center, even when the spreadsheets say otherwise.

The truth is that technology reflects its makers. If the people behind the screen act with care, the products will too.

If they act out of fear or greed, that shows up just as clearly. The code may be invisible, but the values never are. Every button, every policy, every "feature" tells a story about what kind of world we think is worth building.

And maybe that's where nobility in tech starts—not with new rules or frameworks, but with one person deciding that doing good is worth the risk.

Being a Responsible Digital Citizen

People love to blame tech companies. It's easy. They're big, rich, out of reach. But we forget the small stuff we do every day. Every person with a phone or a laptop has a bit of power. Not much, but enough to matter. Every click. Every post. Every time we hit "like" or "share." It all shapes the kind of internet we end up living in.

Most people don't stop to think about that. We just scroll. Tap. Move on. It feels harmless, like tossing a pebble in a lake. But all those pebbles make waves. Some hit harder than we realize.

We scroll. We react. We let algorithms guide us like a current. But being a digital citizen means learning to swim, not just drift.

Thoughtful consumption starts with slowing down. Ask before you share something: who made this, and why? What's it doing to me? If an article or video leaves you angry or smug, that's a sign to pause. Most viral stuff is designed to stir emotion first, facts second. Taking ten seconds to check a source is one small act of resistance. It sounds trivial, but multiplied by millions, it's culture-shifting.

Privacy is another part of digital citizenship that too many people treat like a lost cause. It's not. Protecting your own data isn't paranoia—it's respect, for yourself and for others. Don't forward screenshots without permission. Don't tag people without asking. Don't share what isn't yours to share. We all leave traces online. Being mindful of those traces is part of basic decency now, like closing the door behind you.

Then there's manipulation. Ads, recommendations, outrage cycles—they feed on our impulses. Seeing the tricks doesn't make you bitter. It just clears the fog a bit. You start to notice when something's tugging at your attention, trying to keep you hooked. Then you get to choose. Stay with it or walk away. That's real freedom. Not running from the world, but picking what actually matters to you.

And there's the other part. People. The internet's noisy, sure, but there are still corners where people try to make it better. Small things. Someone leaves a kind comment. Someone helps fix a problem for a stranger. Nobody claps for it, but it counts. Do that. Leave the place a little warmer than you found it.

Digital citizenship isn't a grand moral speech. It's daily practice. It's the moment you think before posting. The time you correct a rumor instead of spreading it. The quiet choice to make the online world a little more human. None of it fixes the system overnight. But it does something that keeps us from becoming the system ourselves.

Building Technology That Deserves Us

Let's be real. The way we build technology right now feels careless. We move fast, break stuff, and call it progress. Code gets written, pushed out, and someone says, "We'll fix it later." But later never really comes, not for the people who get hurt by it. A glitch ruins a small business. A biased system shuts someone out. And the rest of us scroll past it like it's just another headline.

We don't stop to think. Not really. We're so used to speed that slowing down feels wrong, like falling behind. But maybe that's what we need. A pause. A moment to ask the simple, human questions before we ship the next big thing. Who could do this harm? Who's not even in the room when we make these calls?

Right now, those questions barely get asked. The people designing the tools aren't always the ones living with the fallout. You've got engineers and investors making choices that ripple out into classrooms, hospitals, and neighborhoods. The gap between builder and user keeps getting wider, and empathy fades in that space.

If we had any sense, we'd start earlier. Before a single line of code, before the pitch deck. Get regular people involved. The ones who know what it's like to lose a job to automation, or have their data sold, or watch their kid get shaped by an algorithm. Teachers. Caregivers. Students. Folks who live with the consequences every day.

That's what real tech governance should look like. Not just lawyers and CEOs sitting around a table. People who can say,

"Wait, that's not fair," and actually be heard. It's not about stopping progress. It's about steering it with some conscience. When more voices get a say, the results stop being so one-sided. The rules start to make sense for everyone, not just the top few.

We also need to share the benefits better. Technology keeps making a few people rich while leaving others scrambling to catch up. That imbalance isn't progress; it's drift. The tools we build should lift more hands, not just a few. A good system would make sure the gains flow outward—to workers, to local communities, to the next generation who didn't get a say in what we built.

And there's another piece that's harder to measure. Values. Every new system, every algorithm, carries someone's idea of what's right. We can't pretend it's neutral anymore. Machines reflect our choices, our blind spots, our ethics. Governance has to face that head-on. The question isn't just, *can we build it?* But *should we?* And *who does it serve?*

If we get this right, we could build technology that earns our trust again. Something that feels like a partnership instead of a takeover. It won't happen overnight. It'll take patience, honesty, and a lot of uncomfortable conversations. But the alternative is a worse world where machines keep learning faster than our conscience.

Better to slow down a little now than regret the direction later. That's what real stewardship looks like. Not control for control's sake, but care. Building systems that remember what they're for.

Chapter 17
Global Challenges and Noble Solutions

It feels like the world's holding its breath. You can sense it while strolling through the news or stepping outside on a weirdly warm winter morning. Nothing feels steady anymore. Fires where there used to be snow. Streets turning into rivers. The air feels heavier, worn out almost. Climate change isn't some debate now. It's just there; in the dry earth, the missing birds, the smoke that hangs around long after the fire's gone.

And while the planet shifts under us, money keeps pooling in the same few corners. Economic inequality cuts deeper each year. Some people live in worlds built on excess, while others line up at food banks. It's not just unfair—it's corrosive. It eats away at the idea that effort alone can lift you up. You start to see how the rules were written. Who they were written for.

Then there's democracy, wobbling on its old legs. People losing faith, governments bending truth until it snaps. The noise online doesn't help. Everyone shouting, no one listening. Lies travel faster than facts. Leaders play to fear because it works. It's easier to divide than to repair.

Public health was supposed to be a shared concern. Then came the virus. The masks, the panic, the empty shelves. We saw what happens when trust collapses. Systems cracked. Nurses worked until they couldn't stand. People fought over science like it was politics. Even now, the world hasn't really healed. The virus left more than scars; it left suspicion.

And through it all, technology races ahead. Faster than we can think. New tools, new toys, new traps. Jobs vanish overnight. Privacy feels like a rumor. Kids grow up in digital spaces built to keep them from scrolling. Innovation used to feel like hope. Feels like everything's coming at once lately.

You look around and it's all mixed together, heat, money, power, sickness, tech, none of it separate anymore. One thing break, another follows. Drought hits, food runs short, people get angry, and lies spread. It has no distinct beginning or end and is all jumbled.

You then question what you should do about it. On certain days, it's simpler to ignore everything, focus on the now, and live a little existence. However, even that begins to seem off, as if you're sitting motionless as the earth shifts beneath you. The times call for something bigger. Something steadier. A kind of nobility that isn't about titles or pride, but about responsibility. Ordinary people choosing to act with decency when everything around them pulls the other way.

Maybe that's the only real solution left, each of us standing up, however small the step, and saying, *not this time.*

Ground-Level Change:

How Real Progress Begins with People, Not Promises

It's easy to talk about ending poverty or curing disease like their problems waiting for the right genius to fix them. But most progress hasn't come from miracles. It comes from patience, ordinary work. People teach kids to read. Nurses carry medicine down dirt roads. Small groups trying, failing, and trying again.

Take education. In rural Kenya, a group of teachers started something simple—free lunch at school. Attendance shot up. Kids stayed longer, learned better. The ripple effect was real. When one girl finished high school, her younger siblings almost always followed. Education doesn't just open minds; it shifts families, whole villages. And it's not always about textbooks. In Bangladesh, programs teach women to manage small savings groups, turning literacy lessons into small businesses. It's knowledge with purpose.

Healthcare tells the same story. The most effective programs aren't the flashiest ones. In India, a network of community health workers—barefoot doctors, they're called—has saved thousands of lives by doing the basics well. Vaccinations. Prenatal care. Clean water. No fancy gear. No labs filled with machines. Just people showing up, again and again. In Rwanda, after the genocide, they didn't start with grand hospitals or big budgets. They started with broken buildings and a few determined nurses. Clinics rose slowly, sometimes with borrowed tools and hands from the neighborhood. Word spread that help was real, and people started to believe in it. That faith built the rest.

And money—always the hard part. Poverty doesn't fade because someone writes a policy or makes a speech. It changes when people finally have something of their own to hold onto. In some villages across Africa and South Asia, a woman could take out a small loan to buy a goat, or a sewing machine, or a few sacks of grain. Nothing fancy, but enough to stand a little taller. You could see the pride that comes with earning instead of waiting. Charity gives for a day. Ownership changes everything.

In Brazil, they tried something simple but powerful. Families get money if their kids went to school and saw a doctor. It worked. Kids stayed in class. Illnesses dropped. It wasn't a handout; it was a deal built on trust and responsibility. That kind of fairness sticks.

Then there's the safety net—what catches you when the world caves in. Some countries in northern Europe built systems that don't let families fall too far. If you lose your job, you still see a doctor. If your child gets sick, you don't have to choose between medicine and rent. It's not perfect, but it shows what happens when a country decides that care is not a privilege.

Look closely at all these efforts, and you see a pattern. They don't start in conference rooms. They start on the ground. Someone listens. Someone shows up. They work small, and they stay. Change like that doesn't come in slogans. It's slow. It's messy. But it lasts.

Seldom do the people who perform this task garner media attention. They are neighbors, volunteers, teachers, and nurses. Even when everyone else is ignoring them, they continue to appear. Real nobility is there, not in speeches or symbols but in the silent hands that restore what others have abandoned.

Shared Responsibility:

Building a Fairer Global Table

When people talk about global cooperation, it usually sounds clean and easy. A slogan. A shiny sentence in a press release. But it's messy in real life. The kind of messy that smells like coffee gone cold in meeting rooms where no one really listens. You see it in the faces of delegates waiting for their turn to speak, knowing the decisions already made somewhere else. The talk is always about partnership, but the air is thick with ego and suspicion.

And yet, there's no getting away from it anymore. The problems don't stop at borders. Climate, sickness, hunger, data—all of it leaks through walls like smoke. You can't point to one country and say, *That's your problem.* It's everyone's.

Real cooperation doesn't start with handshakes or headlines. It starts with respect, the kind that's quieter than talk. Not the polished kind for cameras, but the kind that listens to even when it's uncomfortable. Every country carries lessons from its own struggle. Small or large, rich or broke, every one of them has something the world could use. The trouble is that too many powerful nations walk into the room thinking they already have the answers. They bring blueprints and funding but forget to bring humility. The people closest to the problem usually see what works best. Listening to them isn't a weakness, it's common sense.

Shared responsibility isn't optional either. The air above us, the water under us, the web between us belongs to one flag. Try to wall it off, and it still seeps through. Pollution, disease, fear—they don't need visas. Sooner or later, the same storm finds everyone. Shared responsibility isn't some lofty moral idea. It's survival, plain and simple. The world's too connected now for one side to rise while the other sinks.

And still, the same voices keep running the show. The same countries, the same faces. They talk about inclusion, but the invites go out after the deal's done. When decisions come from the top, they crumble fast. What looks fair on paper falls apart in the dirt. If the people living with the consequences aren't part of the plan, the plan won't last. Real inclusion isn't about being nice—it's about being smart. The solutions that stick are the ones built by the hands that have to live with them.

It's what makes progress stick.

Then there's fairness, the hardest part. Burdens and benefits must be shared, not hoarded. Countries that grew rich off industry and consumption have a moral duty to lead the clean-up. That's not blame—it's balance. It's not enough to donate to "help." Equity means responsibility scaled to capacity. Those with more resources shoulder more of the load because they can. The same goes for benefits. If a trade deal or development project leaves the poorest still struggling, then it's not cooperation—it's self-interest dressed up in nice words.

And this is where measurement matters. Too many leaders still gauge success by numbers—GDP, market growth, or "strategic wins." Real progress should be measured by who is lifted, not who leads the race. A new trade route means nothing if it leaves fishermen broke. A health initiative fails if the medicine doesn't reach the people who need it most. Before anyone signs a deal or drafts another shiny plan, there's one question that should come first. Does this actually help the people hanging on by a thread? The ones who get hit first and are helped last. If it doesn't, then it's not worth the paper it's printed on.

Global cooperation shouldn't be some pity project. It's not about saving anyone. It's about all of us keeping our heads above water. Real leaders know that. The good ones don't close doors or protect their seats. They pull up more chairs. They make room for voices that usually get drowned out. The quiet ones. The ones that don't talk about everyone else but somehow still make you stop for a second.

If the world worked more like that—just people listening instead of trying to be right all the time—maybe we'd move forward a little. Maybe what we build would last longer. A future that only works for a handful of people isn't much of a future anyway. The real one, the one that matters, has room for everyone. Even the ones we usually forget.

Reimagining Global Systems Through Justice

Unconditional justice sounds like something people say when they want to sound noble. But it's not a dream, not really. It just means fairness shouldn't shift depending on where you happen to be born, or who holds the biggest stick. If the rules only work for the strong, then they're not rules anymore. Their fences built to keep some people out. And we've built too many of those already.

Take trade, for example. That's where you can really see how uneven the ground is. The rich countries talk about free markets like it's a religion, but they play by a different set of rules when it suits them. They guard their own industries, slip in quiet subsidies, and still expect everyone else to stay "open." Poorer countries are told to follow the plan, and when they do, their markets flood with cheap goods they can't compete with.

Local jobs disappear. Families that once managed on little can't even manage that anymore.

If justice meant anything, trade wouldn't be a contest of who can squeeze the most out of the weakest. It would be an exchange built on respect. You sell what you grow, you make what you can, and you get a fair price for it. Not pity. Not leftovers. Just a fair deal. That's all most people ever wanted anyway.

A system that rewards effort and lets small producers stand a chance instead of being crushed by the global machinery.

Financial reform comes next. Right now, the money game is rigged tight. Big banks, credit agencies, and old financial hubs set the tone for everyone else. Debt turns into a weapon. Countries that were supposed to rise just keep paying, year after year, for help that never really helped.

Now flip it for a second—what if the rules were fair? What if interest didn't crush you before you even started? Real debt relief, not another headline. Justice would mean letting people breathe again, rebuilding what's theirs, instead of trapping them in the same old cycle of owing and asking.

Then there's intellectual property. Innovation is supposed to move the world forward, not keep it divided. During health crises, vaccines and medicine shouldn't belong to the highest bidder. Knowledge should travel freely when lives depend on it. Farmers shouldn't lose their seeds to patents. Students shouldn't have to choose between rent and textbooks. Unconditional justice means removing the gatekeepers that profit from scarcity.

Climate agreements might be the hardest test of all. Everyone talks about shared responsibility, but some countries carry more weight while others keep burning. Justice would mean the biggest polluters paying their real share. Not in vague promises or future credits, but in action now. It means helping the nations already knee-deep in floods and droughts, the ones who did least to cause it.

Unconditional justice isn't about perfection. It's about courage. It asks people in power to stop protecting their comfort and start protecting what's left of our shared future. It's messy, slow work, but if the world ever decides to mean what it says about fairness, that's where it begins.

Quiet Courage:

The People Who Keep the World Turning

Sometimes change doesn't start in a meeting room or at the top of a government list. It starts in smaller places. A church basement. A crowded classroom. A dirt road after the rain. The people who move around the world don't always wear suits or speak into microphones. Some of them just keep showing up, day after day, with quiet determination that outlasts the noise.

There's a woman in Kenya named Amina who built a solar cooperative in her village. No fancy degrees, no big investors. Just a group of neighbors who were tired of studying and cooking by kerosene light. They started with one panel, then two, and before long, the whole village was connected. She taught the kids how to wire and maintain the systems themselves. The project spread because people trusted her. They saw it worked. She didn't wait for permission.

In Colombia, a former soldier named Javier started walking into the same towns he once patrolled. Instead of carrying a gun, he carried books and sports gear. He started with nothing much. Just a dusty room behind an old church. A few chairs, a cracked window, and a coffee pot that barely worked. Kids began showing up after school, the kind of kids people crossed the street to avoid. He didn't preach or make promises. Just kept the door open and said, "You can sit here if you want." Some did. Most didn't, at first. But word got around that it was safe.

He'd been in trouble once too. He knew what that edge felt like. He used to say he couldn't change what had already happened, but maybe he could keep someone else from falling the same way.

At first, the kids came just to pass the time. Some stayed. They swept the floor, patched old chairs, and kept the lights on when the bills piled up.

One came back years later, with a badge around his neck, tired eyes, but smiling. He said this place was the first time he ever felt worth something. The man didn't say much. Just nodded and looked around, like the room held the proof.

Leila's story moves quieter. She grew up between tents and tall buildings, never in one place long enough to unpack. Learned to listen more than speak. Now she works in those big meeting rooms where people wear suits and talk in circles. She doesn't fight for the loudest word. She waits for the moment when someone finally drops the act. Then she finds a way through. She says peace doesn't come from signatures. It comes from people who refuse to stop talking to each other.

None of these people look like heroes. They don't make headlines. They just keep showing up. They do what they can, even when it feels like it won't matter. That's the part that gets to me—the steady, quiet trying. It's not about fame or winning. It's about not giving up on what still might work.

If noble leadership means anything, it's probably this. A hand that doesn't pull away. A voice that stays calm when everyone else is shouting. Small steps repeated until something shifts. The world changes that way, inch by inch. Not because someone demanded it, but because a few people never stopped believing it could.

Chapter 18
The Future of Nobility: A Call to Action

It's easy to picture a future packed with machines that outthink us. What's harder is picturing one where we still hold on to what matters. The screens will keep getting sharper, the code smoother, the systems quicker. But that's not the real test. The real test is us, what we choose to build, what we choose to let go.

Picture it for a moment. A world where AI doesn't replace judgment but sharpens it. Where data helps us see clearly, not hide behind numbers. Where algorithms assist doctors in saving lives instead of selling ads. Where education uses technology to open minds, not close them into echo chambers. That kind of balance isn't automatic. It must be built, piece by piece, by people who care more about purpose than profit.

Progress without any sense of value burns out fast. You can see it now in the tired faces, the endless updates, the quiet pressure to keep up. We're all plugged in, yet most days still feel alone. Maybe the next real breakthrough isn't faster tech at all. Maybe it's learning when to stop, when to ask the hard question: should we make it, just because we can?

Imagine organizations run by leaders who still look others in the eye before making decisions that affect millions. Places where success isn't just measured in quarterly gains but in how many lives were made better, how much harm was avoided. That's not naïve. It's possible. It starts small, in the choices of designers, teachers, investors, and voters. In the quiet moment before someone hits "publish" or "approve."

AI could become a mirror for what's best in us if we teach it that way. It learns from what we feed it. If we give it greed, it will reflect that. If we give it compassion, it might help us remember our own. The challenge isn't about control. It's about character.

A future worth having will need both intelligence and humility. Machines might calculate, but only humans can care. Maybe that's where nobility belongs now—not to titles or power, but in how we hold technology to a higher purpose. The world ahead could be brighter than we expect, but only if we bring our best selves with us.

Living Nobly in Every Sphere of Life

Nobility doesn't start on stage. It begins quietly, in the tiny corners of your day. How you talk to yourself in the morning. The patience you show when no one's there to notice. The moment you stop and take a breath instead of firing back. It's not the kind of thing anyone claps for. But those small choices start to dig in. They take root. And before you know it, they've shaped who you are.

Maybe it's writing a few lines in a journal. Maybe it's stepping outside instead of grabbing your phone. Maybe it's just breathing through the noise. That's the real work — the quiet kind that builds strength from the inside out.

Then there's family, the real testing ground for any virtue. It's easy to speak about kindness in theory. It is harder to live with the people who know your worst days. Nobility at home isn't perfect, it's honesty. It's saying sorry without defensiveness, listening even when you're tired, setting an example instead of giving speeches. Kids, friends, partners, they see what you do more than what you say. Every small act plants something.

Workplaces are another field entirely. The temptation to stay quiet is constant. The project goes wrong. Someone takes credit that's not theirs. You notice something that feels off. Most of the time, noble leadership isn't about big speeches or spotlight moments. It's the quiet decision to do what's fair when it costs you a little. Giving credit when it's due. Saying the hard truth instead of what's easy. Drawing a line when everyone else shrugs. Change starts that way — one person deciding not to look away.

Then there's the community around you — the block, the people next door. The real-world shifts into small rooms long before they ever do on screens. Helping out. Mentoring a kid. Checking on someone who's been alone too long. None of it looks dramatic, but it adds up. You don't have to fix the world. Just handle your corner with care. Sometimes that's enough to lift the whole street.

And finally, there's the civic side. The part that reaches beyond comfort and asks you to care about systems. Read the news from more than one source. Ask questions before sharing a headline. Vote, even when it feels pointless. Support people and policies that put integrity before image.

These five spaces—self, family, work, community, country—they overlap and feed each other. You grow in one, you get stronger in the rest. It's not about being perfect. It's about staying awake. Doing the next thing right even when it's small. That's how nobility survives. Quietly. Day by day.

Living the Small Truths

You wake up, sit there, stare at the light. Think about the day before. Not everything went right, not everything went wrong. Just life. You tell yourself, *Alright, today I'll try again.* Maybe be kinder. Maybe listen more. Maybe just breathe before reacting. That's all.

Occasionally, you stop thinking about yourself. You notice someone else. A neighbor, tired, struggling with groceries. A coworker who looks worn out but still smiles anyway. You help, or at least you try. It doesn't fix the world, but it makes the air a bit lighter.

Then maybe once a week, you do something that connects you. Not big. Not perfect. Just humans. Visit an older friend. Call someone you've ignored. Drop off food, fix something, clean up after others. Not for credit. Just to remind yourself that you're not the only one trying.

Each month, pick up something new to learn. A subject you know nothing about. Something that shakes your head open. You'll hate it at first, maybe get frustrated, maybe feel dumb. That's good. It means you're not coasting.

And once a year, you give something away. It could be money, could be time, could be yourself. You don't post it.

You don't talk about it. You just do it and move on. Quiet things like that, they build a kind of peace you can't fake.

You'll forget all this sometimes. Drift off. Lose patience. Everyone does. Then something small brings you back. A story. A memory. Someone's eyes. You remember who you wanted to be. And you try again.

That's what noble living really is. No big speeches. No perfect plan. Just ordinary days done with a bit more heart.

From Inner Work to Outer Change

It starts small, always does. A few honest moments with yourself. The quiet kind that no one sees. You notice the corners you cut, the times you looked away, the little ways comfort won over conscience. It's not guilt, exactly. More like a tug. A voice saying, *Alright, time to do better.*

The shift happens slowly. You catch yourself holding back a harsh word. You ask a question before you judge. You listen all the way through instead of waiting for your turn to talk. None of them feels heroic. It's just lighter, somehow, cleaner. The kind of thing you'd never brag about because you know it's only the beginning.

Then something changes. The inner work starts to spill outward. You start noticing things you used to scroll past. People sleep outside office buildings. A coworker who keeps showing up tired. Someone always left out of the group chat. It hits differently when you're awake to it. The world stops being background noise and starts becoming personal.

So, you do something, even if it's small. You speak up in a meeting when silence feels easier. You offer help without making a show of it. You write to someone you've wronged. That's how it builds—quiet acts stacking up until you realize you've been living differently for a while now.

The truth is no one changes the world in one clean move. It never happens clean. It's slow, awkward, sometimes messy as hell. You'll have a good day, maybe two, then trip over the same old habit and wonder if you've learned anything at all. But you keep at it. Something inside just won't go quiet.

After a while, the line between what's inside you and what's outside starts to fade. What you think becomes what you do. What you do changes what you think. It loops around, over and over. Not perfect. Just steady.

People pick it up before you even say a word. A shift in tone. A softer look. The way a room feels different for no clear reason. Nobody posts about it or gives it a name, but it spreads anyway. That's how real change moves. Quiet. Uneven. Mostly unseen.

Then someone else catches it. They start in their own way. Maybe they help without being asked. Maybe they listen longer. You notice, and it hits you — this is how it happens. Regular people, doing what they can, no big banners, no permission slips. Just care, passed from one hand to the next.

You stop thinking about saving the world. You just stop pretending it's fine when it's not. That's the shift. The small things build muscle over time. You get stronger at noticing what matters. You stay put when it's easier to look away. And somehow, that steadiness starts to move the ground a little.

So, you keep at it. Even when no one sees, even when it feels like nothing's changing. Because the things no one sees are usually the ones holding everything up.

Holding Each Other Up

Doing good things alone is hard. Keeping it going is even harder. That's why people need each other. Not crowds. Just a few who show up when it matters.

A community doesn't have to be big or loud. It can start in a living room or a coffee shop. A few people meet once a month. They talk about what they tried, what fell apart, what sort of worked before life knocked it over again. There's some laughter, some frustration. Then they head back out and give it another shot. That's how real change happens — not through big plans, but through small groups that keep you steady when everything else pulls you off track.

Some call it accountability. Some just call it friendship. The label doesn't matter. What matters is honesty. Someone who asks, "Hey, did you actually do it?" without turning it into a lecture. Someone who reminds you of what you said you wanted when you're ready to walk away. Alone, it's easy to come up with reasons. Together, it's harder to hide.

It's harder when someone cares enough to ask.

Learning circles works the same way. People come together with questions instead of answers. They trade stories, share books, and swap mistakes. Nobody's trying to be the expert.

Everyone's just trying to grow without losing their soul in the process. You learn faster when you're not pretending to know everything.

Then there are service collectives. Loose groups that decide to do something, not just talk. Maybe they clean up a park, or tutor kids, or help a neighbor's small business stay open. They don't wait for permission. They see a gap and fill it. The work isn't glamorous. It's usually messy and unplanned. But it builds a kind of bond that can't be faked.

Mentoring is another piece of it. Not the formal kind with polished speeches and coffee schedules. Real mentoring happens in passing moments. A quick call. A quiet word. A person who's been through the fire saying, "You'll get through it too." It's how wisdom survives.

These small communities keep noble action alive when everything else feels too big or too cynical. They remind you that one good act still matters. That you're not crazy for trying to live by principles that the world sometimes laughs at.

In the end, staying noble isn't about willpower. It's about connections. About not letting the noise drown out what's good. About finding your people and holding on to them, even when the current runs the other way.

One Soul, Many Lives

There's this old line from Saadi Shirazi — "Human beings are members of a whole." You may or may not have heard it. It feels new even though he penned it hundreds of years ago.

According to him, even if they pretend not to, everyone else feels pain when one person does. Truth like that never gets old.

We forget that most days. We move through life like everyone's separate, like what happens to someone else can't reach us. Then something cracks — a headline, a neighbor having a hard time, a small moment that reminds you how close everything actually is. And it hits. You feel it, not in your head, but right in your chest. That's what Saadi meant. He wasn't being poetic for the sake of it. He lived through war, loss, and famine. He saw people turn cold and still said, *No, we're connected.* That's what makes his words heavy. He didn't write them from comfort. He wrote them from the edge.

If one person's pain echoes through everyone, then maybe every small act matters more than we think. Holding a door. Listening when someone's talking. Standing up when something's wrong, even if your voice shakes. It sounds simple until you try doing it every day.

The world keeps trying to split us apart. It sells a division like a product. Rich versus poor, young versus old, one country against another. But nobility, real nobility, starts when someone says, "No, we're still part of the same thing." It's not about saving anyone. It's about remembering what's already true.

Saadi's words still hang in the air because they don't belong to one place or faith. They're a reminder, not a rule. You don't have to change the world. Just don't harden. Keep the door open. Let someone else's pain reach you for a second. That's where everything good will start.

So maybe that's the call. Not a grand one, just a quiet one. Pay attention. Feel something. Do one small thing that makes the world less cruel. It doesn't have to be fancy, just real. The kind of reality that lasts longer than all the noise.

Conclusion

The journey of knowing, recognizing, pinpointing, educating, and shaping a Noble ended in this book, of course, with my limited knowledge, understanding, and comprehension of relations between AI and NI.

As much as being a Nobel sounds and feels easy, at the same time, the amount of dedication needed to stay on the path is not. It requires passion, forgiveness, and looking up to meaningful values in life, particularly life at large.

Every one of us in some ways must feel committed to elevate the message of nobility; some by teaching others, some by acting on their daily routines, some by learning to give hands, some at home teaching their kids and children, some by spreading the word of nobility through books, audio books, radio, television, social medias podcasts, and others by putting efforts to input practical examples and introducing steps for this valuable agenda.

Be a Nobel!

Pershia Dashti

www.ingramcontent.com/pod-product-compliance
Lightning Source LLC
LaVergne TN
LVHW011520200926
848484LV00001B/86